Sleep Disorders in Children

Anna Wani • Imran S. Khawaja
Editors

Sleep Disorders in Children

A Primer for Primary Care Providers

Editors
Anna Wani
Family and Community Medicine,
and Pediatrics
The University of Texas Southwestern
Medical Center
Plano, TX, USA

Imran S. Khawaja
Psychiatry and Neurology
The University of Texas Southwestern
Medical Center
Dallas, TX, USA

ISBN 978-3-031-92165-0   ISBN 978-3-031-92166-7   (eBook)
https://doi.org/10.1007/978-3-031-92166-7

© The Editor(s) (if applicable) and The Author(s), under exclusive license to Springer Nature Switzerland AG 2025, corrected publication 2025

This work is subject to copyright. All rights are solely and exclusively licensed by the Publisher, whether the whole or part of the material is concerned, specifically the rights of translation, reprinting, reuse of illustrations, recitation, broadcasting, reproduction on microfilms or in any other physical way, and transmission or information storage and retrieval, electronic adaptation, computer software, or by similar or dissimilar methodology now known or hereafter developed.

The use of general descriptive names, registered names, trademarks, service marks, etc. in this publication does not imply, even in the absence of a specific statement, that such names are exempt from the relevant protective laws and regulations and therefore free for general use.

The publisher, the authors and the editors are safe to assume that the advice and information in this book are believed to be true and accurate at the date of publication. Neither the publisher nor the authors or the editors give a warranty, expressed or implied, with respect to the material contained herein or for any errors or omissions that may have been made. The publisher remains neutral with regard to jurisdictional claims in published maps and institutional affiliations.

This Springer imprint is published by the registered company Springer Nature Switzerland AG
The registered company address is: Gewerbestrasse 11, 6330 Cham, Switzerland

If disposing of this product, please recycle the paper.

To my children,

I love you, I'm proud to be your mama.

To my husband,

for your boundless patience and unwavering love.

To my parents,

for sacrificing your todays for our tomorrows.

To my brother and sister,

for keeping me humble and your unbridled honesty.

And to God,

through whom it is all possible.

Preface

Sleep is a fundamental part of life for everyone—whether it's the quiet rest of an infant, the deep sleep of a toddler, or the restorative slumber of an adolescent. As a primary care practitioner, you understand that sleep is essential for health, development, and well-being. Just as you, yourself, sleep each night, so too do your patients, and it's in those hours of rest that much of their physical and mental growth occurs. This book is designed to help you recognize and address sleep-related concerns across all ages, from infancy to adolescence, and to guide families through the complexities of sleep disorders. Understanding the importance of sleep and its role in overall health will empower you to make a lasting impact on the well-being of your patients, as well as deepen your ability to care for them holistically.

This book, *Pediatric Sleep Medicine: A Clinical Reference for the Primary Care Practitioner*, is designed to serve as a comprehensive guide for healthcare providers who are navigating the multifaceted world of pediatric sleep. Its purpose is to provide evidence-based knowledge and practical insights to help you identify, diagnose, and treat sleep disturbances in children of all ages.

As Maya Angelou so eloquently stated, "When you get, give. When you learn, teach." This book is the result of years of research, clinical experience, and collaboration with experts in pediatric sleep medicine. I hope the knowledge shared here empowers you to provide better care for your young patients, while also encouraging you to pass on what you learn to others in your practice and community.

The information contained within this book is intended to be a resource for you—a tool to help you navigate the intricacies of pediatric sleep and to ultimately ensure the well-being of the children you serve. I hope it serves as both a reference and a reminder of the profound impact that sleep has on every aspect of a child's life.

Thank you for your commitment to improving the health of children through thoughtful care, ongoing learning, and, most importantly, giving back to the families who depend on your expertise.

Plano, TX, USA	Anna Wani
Dallas, TX, USA	Imran S. Khawaja

Acknowledgments

I am profoundly grateful to the insightful contributors of each chapter, who generously dedicated their time and expertise to bring this book to life. Your commitment and knowledge have made this resource possible. I also want to express my deep appreciation to my mentors, whose guidance and wisdom have continually shaped my journey. Lastly, to my patients—thank you for being my greatest teachers. Your trust, stories, and experiences have been a constant source of learning and inspiration. May your dreams be filled with promise and peace.

Contents

Part I The Importance of Sleep

1. **Sleep Deprivation in Children and Adolescents** 3
   Lissette Jimenez, Christine Y. Liu, and Aslee Sierra

2. **Screen Time and Sleep** 17
   Sullafa Kadura and Poulomee Tripathi

3. **Sleep, Chronotype, and Learning: A Developmental Perspective** 23
   Marty Martin and Zeeshawn Malik

4. **School Start Times** 33
   Tanya Martinez-Fernandez, Rachel Manuel, Zaiba Jetpuri, and Swetha Gannarapu

Part II Sleep Through the Years

5. **Infant Sleep** 43
   Likhita Shaik, Daniel Rongo, Janey Dudley, Humza Siddiqi, and Anna Wani

6. **Toddler Sleep** 51
   Liz Lezama, Shan Luong, Victoria Udezi, Esha Banwatt, and Elisa Basora-Rovira

7. **Sleep in School-Aged Children** 63
   Janey Dudley, Likhita Shaik, and Shan Luong

8. **Sleep in Adolescents** 71
   Crystal Cassimere and Michelle Caraballo

Part III Common Sleep Disorders

9. **Obstructive Sleep Apnea** 85
   Aarti Shakkottai and Manahil Firdaus

10. **Other Sleep Apneas** 95
    Janey Dudley, Daniel Rongo, and Brittney Pryor Craig

11 **Parasomnias in Children** 101
    David Rongo, Daniel Rongo, and Shan Luong

12 **Sleep-Related Movement Disorders** 111
    Lourdes M. DelRosso

13 **Excessive Daytime Sleepiness in Pediatric Age** 131
    Aysegul Karaca, Maria Rodriguez, Hadia Tahir,
    and Syed Kamal Naqvi

Part IV Sleep in Special Populations

14 **Neurological Disorders** 147
    Daniel Rongo, Rabab Naqvi, Shan Luong, and Sejal V. Jain

15 **The Restless Mind: Sleep Disorders and Mental
    Health in Children** 161
    Imran S. Khawaja and Joshua Robbins

16 **Sleep in Special Populations: Genetic Syndromes** 167
    Likhita Shaik, Janey Dudley, Daniel Rongo, Elisa Basora,
    and Anna Wani

**Correction to: Sleep, Chronotype, and Learning:
A Developmental Perspective** C1
Marty Martin and Zeeshawn Malik

Index 177

Part I
The Importance of Sleep

Sleep Deprivation in Children and Adolescents

Lissette Jimenez, Christine Y. Liu, and Aslee Sierra

Introduction

Sleep deprivation remains an under recognized public health concern [1]. Despite established sleep duration guidelines, most children and adolescents fail to meet recommended sleep requirements for optimal health [2, 3]. Short sleep duration—defined as less than 9 h for children aged 6–12 years and less than 8 h for teens aged 13–18 years—is highly prevalent, with 57.8% of middle school students and 72.7% of high school students not getting enough sleep [4]. In today's fast-paced world, sleep is frequently de-prioritized, often sacrificed in favor of academic demands, extracurricular activities, screen time, and social obligations. Numerous societal and personal pressures compete with sleep and contribute to its erosion. Health disparities further compound the issue in underserved populations. In addition to sleep quantity, other critical yet often overlooked aspects of sleep health include sleep timing (circadian alignment), sleep quality (depth and continuity), napping patterns, and individual variability in sleep needs [5, 6]. There is growing evidence that short sleep duration adversely affects cognitive, emotional, and cardiometabolic health in children and adolescents. Adequate sleep is essential for optimal functioning across all age groups, but it plays an especially crucial role in children

L. Jimenez (✉)
Department of Medicine, Neurology and Sleep Medicine Section, VA Caribbean Health Care System, San Juan, PR, USA
e-mail: lissette.jimenez-davila@va.gov

C. Y. Liu
Department of Family and Community Medicine, UT Southwestern Medical Center, Dallas, TX, USA
e-mail: christine.kwok@utsouthwestern.edu

A. Sierra
Department of Internal Medicine, VA Caribbean Health Care System, San Juan, PR, USA
e-mail: aslee.sierra-gonzalez@va.gov

and teens, as it supports their physical, emotional, and cognitive development [7, 8]. Poor sleep in youth also impacts caregiver wellbeing, increasing stress and impairing daily functioning [9, 10]. Promoting sleep health requires coordinated prevention, identification, and intervention efforts across families, healthcare systems, and society at large. Sleep must be valued and protected as a foundational pillar of overall health.

Sleep Needs, Causes, and Risk Factors for Sleep Deprivation

Sleep and wakefulness are regulated by interaction of the homeostatic and circadian processes. The homeostatic sleep pressure (Process S) increases with wakefulness throughout the day. The circadian rhythm (Process C) aligns sleep and many physiologic processes with the environmental light and dark cycles. The hypothalamic suprachiasmatic nucleus (SCN) is the central pacemaker and produces an alerting signal that promotes daytime wakefulness [11, 12].

Neonates sleep 60–70% of the day and have irregular sleep until about 10–12 weeks of age when the circadian rhythm emerges. Sleep homeostatic pressure is greater in children, and they relieve it with naps. Children nap until about 5 years of age, and the amount of required sleep continues to decrease with time, with the fastest decline in the first year of age [13, 14]. The American Academy of Sleep Medicine recommends sleep duration of 12–16 hours for infants 4–12 months, 11–14 hours for toddlers 1–2 years, 10–13 hours for preschool children 3–5 years, 9–12 hours for school-aged children 6–12 years, and 8–10 hours for teenagers 14–16 years [2]. Recovery after sleep deprivation may take longer in the young. Young adults were found to be sleepier and have greater slowing in response times than healthy older adults to acute sleep deprivation [15]. It is important to mention that sleep duration recommendations should be individualized. Optimal sleep duration may have a dose-response and may differ between outcomes (physical, cognitive, behavioral) [16] (Table 1.1).

Table 1.1 Recommended hours of sleep per age

Age	Hours of Sleep Recommended
4 – 12 months	14 – 16
1 – 2 years	11 – 14
3 – 4 years	10 – 13
6 – 12 years	9 – 12

Data source from Paruthi S, Brooks LJ, D'Ambrosio C, Hall WA, Kotagal S, Lloyd RM, et al

Sleep deprivation is more aggravating in adolescence due to physical and exogenous factors. Pubertal hormonal changes, including a delay in melatonin secretion, cause a physiologic delay in sleep time that aggravates sleep debt when coupled with early school start times. Circadian rhythms influence alertness and cognitive performance. Growth hormones are secreted during slow wave sleep. Changes in testosterone and estradiol have been linked to changes in sleep patterns [7]. Blue light from screens reduces sleep-promoting melatonin production [17]. External pressures from school assignments, extracurricular activities, limited exposure to natural light during the day and exposure to artificial light at night, lack of physical activity and screen time are related to delayed initiation of sleep and less restful slumber [18–20].

Delay in school start times has been recommended by multiple groups since it is associated with improved sleep duration, alertness, mood, health and academic performance [21, 22].

There are documented differences in sleep duration by race/ethnicity and socioeconomic status. Black children were reported to sleep 34 min less per night than White children and go to bed on average 53 min later than White children [23]. Hispanic children ages 6–12 living below the poverty threshold faced nearly twice the odds of not meeting sleep duration recommendations [24].

Diagnosis and Assessment

Parents, teachers, and health professionals often underappreciate the prevalence of sleep deprivation. Adolescent reports often overestimate actual sleep (Short 2012). Signs of insufficient sleep may be normalized or misidentified as a mood or behavioral disorder [25, 26].

Sleep deprivation results from getting less sleep than needed for proper health and functioning. Acute sleep deprivation involves little to no sleep or a reduction in usual sleep for 1–2 days, often extending wakefulness beyond 16–18 hours. In chronic sleep deprivation, there is insufficient sleep almost daily for at least 3 months and is characterized by persistent daytime sleepiness [27].

The clinical evaluation should include a detailed sleep history from the patient, family member, and if possible, from school. Sleep duration and continuity, napping, and daytime functioning should be assessed. Sleep diaries and objective sleep patterns scales, such as the *Sleep Disturbances Scale for Children* and the *Sleep Disorders Inventory for Students—Children and Adolescents* are useful tools for obtaining valuable information on sleeping patterns in pediatrics [28]. Sleep deprivation may present as behavior issues, mood dysregulation, and cognitive problems. Fatigue, daytime sleepiness, hyperactivity, inattention, impaired school performance, substance abuse, and risk-taking behaviors may all represent insufficient sleep. Due to the close link between sleep and emotions, all patients with mood complaints must be screened for insufficient sleep.

Not-to-be-missed signs of insufficient sleep include trouble waking up and getting out of bed, long afternoon naps and weekend sleep times 2 hours or longer than during school nights [5, 25] (Fig. 1.1).

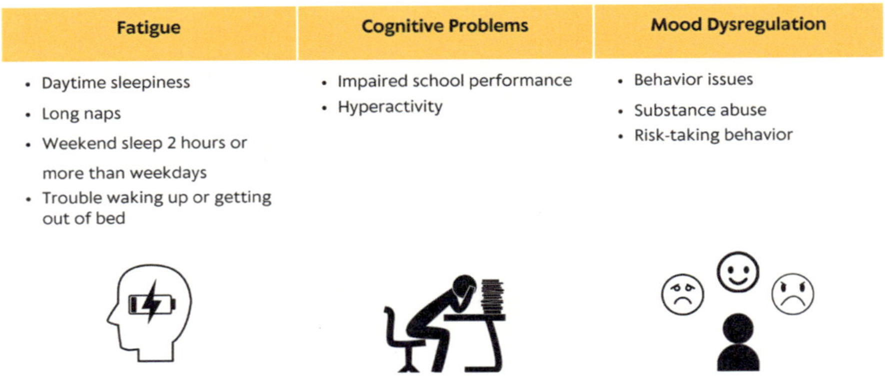

Fatigue	Cognitive Problems	Mood Dysregulation
• Daytime sleepiness • Long naps • Weekend sleep 2 hours or more than weekdays • Trouble waking up or getting out of bed	• Impaired school performance • Hyperactivity	• Behavior issues • Substance abuse • Risk-taking behavior

Fig. 1.1 Signs of potential sleep derivation

The physical exam is usually normal in sleep deprivation and, if abnormal, should prompt exploration of other causes of the presenting symptoms, for example, sleep apnea and periodic leg movement disorder. Sleep deprivation is diagnosed by history rather than formal sleep studies. Nevertheless, polysomnography or actigraphy may be necessary for cases of suspecting the existence of a sleep disorder or when associated abnormalities are suspected [29].

Consequences of Sleep Deprivation: What Happens If You Don't Get Enough Sleep?

Developmental Impacts, Cognitive Impairments, Cognitive Functioning, Academic Performance, and Behavior Effects

It is crucial for brain development to have an adequate amount of sleep. Sleep has an important role in synaptic homeostasis, cortical plasticity, memory, and learning. REM sleep strengthens and maintains newly formed synapses [30]. Sleep in infancy has been correlated with cognitive performance during preschool [31]. Sleep deprivation is considered to have negative effects on cognitive and emotional development, which may result in long-term attention, learning, and emotion regulation deficits [32–37]. Children with sleep disturbances had smaller grey matter volumes and thinner prefrontal cortex [38].

Sleep deprivation significantly impairs executive functions such as attention and memory [39]. Research has shown that sleep deprivation impairs academic performance, and it also predisposes for behavioral problems. Chronic sleep deprivation has been associated with increased risk for mental health problems and with higher levels of aggression and hyperactivity [40–42]. Studies have shown that adolescents who experienced shorter sleep duration and went to bed later during the night during their childhood and teenage years were at a higher risk of alcohol use or trying marijuana by the age of 15. Ensuring adequate sleep through childhood may help lower the risk of substance use during adolescence [42, 43].

A chronic reduction of just 1 hour of sleep per night in early childhood had an adverse effect in the child's cognitive performance at school entry. Short sleep was associated with hyperactivity-impulsivity and lower cognitive performance on neurodevelopmental tests [44].

Sleep deprivation has an adverse impact on cognitive functions such as attention, memory, executive function, etc. Adolescent sleep deprivation is associated with greater difficulty in challenging problem-solving dispositions and relatively poor academic accomplishment. Lower academic performance is associated with poorer grades and increased absenteeism from school [39–42, 45].

Sports injuries are more common in children who are sleep deprived.

Sleep and Cardiometabolic Health

Short sleep duration is associated with obesity in children and adolescents. Cross-sectional studies from around the world correlate short sleep and obesity in children and adults [46–49]. Poor sleep quality, independent of duration, was also associated with obesity [50]. Although limited, intervention studies increasing sleep duration have shown improvements on BMI and obesity [49].

Sleep is a fundamental aspect of health that plays a vital role in various physiological processes, including metabolism, cardiovascular function, and immune response. In recent years, growing evidence has linked sleep duration and quality to cardiometabolic health, highlighting the importance of adequate sleep in reducing the risk of cardiovascular disease, type 2 diabetes, obesity, and other metabolic disorders [50, 51].

The regulation of metabolism and cardiovascular function is intricately connected to sleep patterns. Insufficient or poor-quality sleep has been shown to disrupt endocrine and metabolic processes, leading to adverse outcomes such as an increase in blood pressure, dysregulated glucose metabolism, and altered lipid profiles, all of which are key contributors to cardiometabolic disease. This disruption is often medicated through mechanisms like heightened sympathetic nervous system and insulin resistance [50, 51].

Recent studies have found that shorter sleep durations are correlated with increased body mass index (BMI), elevated blood pressures, and adverse lipid profiles in youth. The implications of these findings are significant, as they suggest that inadequate sleep during childhood can set the stage for chronic health problems in adulthood, including cardiovascular disease and metabolic syndrome [51].

Management and Treatment

Prevention Strategies

It is important to educate families and patients on sleep hygiene, its importance, and its impact on overall health. Parent and primary care physicians should provide guidance on sleep hygiene practices, as well as information concerning the ways lifestyle choices impact upon sleep [45].

Bedtime Routine

Establishing and maintaining a consistent bedtime routine is crucial for healthy sleep. Children with irregular bedtimes were found to have a greater likelihood of having behavioral problems, and children and adolescents were also found to have (Pressman and Imber 2011; Komada 2011). Parental behavior was found to predict sleep duration in children (Mindell 2006). There are behavioral interventions which may effectively manage sleep deprivation, such as cognitive-behavioral therapy for insomnia (CBT-I). These include stimulus control, sleep restriction, and cognitive restructuring [52]. In addition to this, it is crucial to provide patients and their families with information about sleep hygiene (i.e., adhering to a schedule or avoiding blue light devices), as well as lifestyle advice, including establishing an appropriate sleeping environment for relaxation [53].

Ways to enhance the amount and quality of sleep often involve [53, 54] addressing comorbid conditions like ADHD or anxiety, which can lead to improvement with treatment of the underlying issue, such as behavioral therapy. This may include obtaining mental health consultations and delivering specific interventions [55]. Pharmacological treatment should be considered only when behavioral interventions have failed. Medications such as melatonin, antihistamine, and hypnotics may be prescribed but there is limited evidence to determine its efficacy and safety of their use in the long-term basis [56] (Fig. 1.2).

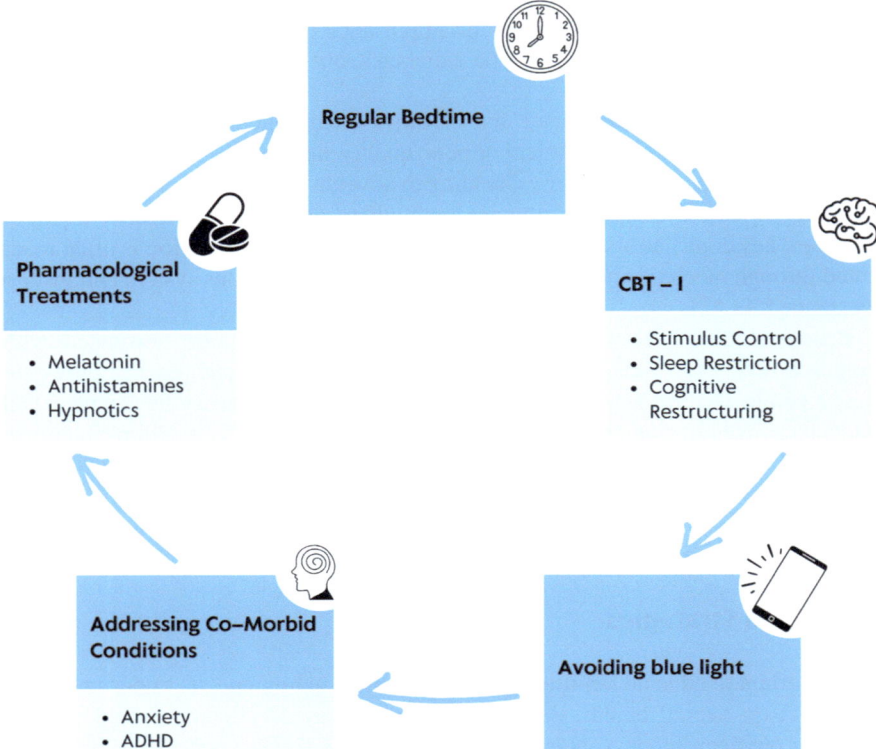

Fig. 1.2 Treatment for sleep deprivation

Special Considerations

Children with developmental disabilities, such as children with autism spectrum disorder, are a growing sub-population that often require individualized sleep interventions. They might also involve targeted bedtime routines, accommodations for sensory issues, and personalized behavioral therapies [55, 56].

Conclusion

Sleep is a multifaceted process that is regulated by the central nervous system, relying on neurotransmitters to regulate the different stages of sleep. REM sleep is a stage for memory consolidation and emotional processing while non-REM sleep stages are associated with physical repair and growth in humans. However, sleep deprivation disrupts them and ends up affecting neurological development and emotional regulation in a well-established scenario [57]. Without sleep, our children cannot grow or develop physically and cognitively due to the disruption in the release of this hormone while we sleep.

The National Sleep Foundation demonstrates that children 6–13 years should receive anywhere from 9–11 hours of sleep every night while adolescents aged between 14 to 16 need about 8 to 10. Nonetheless, recent studies have shown that many children and teenagers are not in compliance with these recommended sleep durations. Approximately 70% of adolescents report insufficient sleep during weekdays [37]. The reason is simple: more hours on screens, academic pressures, and social activities.

Missing a night's rest can disrupt the balance between neurotransmitters and hormones responsible for promoting wakefulness and sleep promotion. Chronic sleep deprivation disrupts the hypothalamic-pituitary-adrenal (HPA) axis, which causes elevated cortisol levels and subsequent mood and cognitive impairment as well.

Mental health issues may manifest as sleep disturbances in children and adolescents with anxiety or depression. These can result in not being able to fall asleep, multiple night wakings, and early morning awakenings [58]. Sleep and mood disorders are interrelated in a two-way manner; sleep can independently worsen psychiatric symptoms, but conversely mental health conditions could lead to changes in the pattern of impaired functioning related to circadian rhythms [59]. An erratic daytime schedule and differing sleep times can blur the body's internal to time, making it harder to get a decent night peak physiological condition [57].

Case

CC: Decreased school performance, Mom wants ADD/ADHD assessment.

HPI: JR is a 12-year-old boy, who presents to clinic with his mother. Mother reports JR has had decreased school performance in 6th-grade, his first year in middle school. He received some B's and one C his first semester, even though he previously was a straight-A student. He is more fidgety in the morning to the point that he has had to be separated from his peers to the back so as not to distract other students. But in the afternoons, his teachers catch him staring off into space and have to reorient him constantly to stay on task.

When asked how he is feeling, JR says he is "fine" and denies any symptoms.

Sleep History

JR usually goes to sleep around 10 pm, after his homework is done. He used to eat dinner after sports, do his homework in 30 minutes. But now his homework takes 1–2 hours, pushing his sleep time later. Mom is perplexed about his homework, saying, "It doesn't appear that hard, but he seems too distracted to get it done faster." Mom allows him iPad game time for 30 minutes prior to bed to unwind. Even though he appears tired at bedtime, it now takes him 30 minutes or more to fall asleep, longer than in the past. He does not wake up at night and denies nightmares. He sets his alarm for 6 AM, but he is hard to awaken in the morning for school, usually taking parents 20–30 minutes of repeated visits to his room to make him get up. He sleeps on average 7.5 hours a night on weekdays, 8 hours on weekends. He could set his alarm for 6:30, but parents are concerned that would run him later. JR does not have time to nap, and when asked says he doesn't feel like napping. Mom states he doesn't have time to nap long but does nap in the car to school, from school, and sometimes on the weekends between events; he is easy to rouse from naps. Mom states she does not hear JR snoring and has never witnessed any apnea. He does not have jaw pain or dry mouth upon awakening.

Past Medical History
- A broken arm when he was 5, recovered with casting
- Allergic rhinitis

Surgical History: none

Social History

JR plays soccer after school 3 days a week and also at Saturday tournaments, which consist of 1–2 games. He is one of the star players on his team and loves soccer. He also does Taekwondo the other 2 days of the week after school and enjoys that. He goes to church Sunday morning and then takes piano lessons on Sunday afternoon; he states he enjoys this less but doesn't mind it, just dislikes 30 minutes of practice daily. He has many friends at school but has less time to hang out with friends after school than before. He gets along with most people.

JR lives at home with Dad, Mom, 2 older sisters.

Mom was asked to step out. JR denied any alcohol, tobacco/vaping, drugs. He vehemently denies any romantic relationships. Denies any abuse or bullying. JR states parents have parental controls set up on his iPad and makes him leave it outside when he goes to sleep.

Mom was brought back in for the patient's exam.

Medication: Claritin 10 mg by mouth daily.

NKDA

Review of Systems:
- General: no fever or night sweats, weight loss/gain significantly
ENT: no congestion, rhinorrhea, dry mouth, nor sore throat
CV: no palpitations, chest pain
Pulmonary: no shortness of breath, wheezing, nor coughing
GI: no nausea, vomiting, diarrhea, nor constipation
Neurological: occasional headaches in the afternoon (1 every 2–3 weeks), no vision changes nor dizziness
Psychiatric: no depression nor anxiety
MSK: no joint pain nor joint swelling, no jaw tightness
Skin: no rashes or itching

Exam:
- Vital Signs Stable
General: no apparent distress
ENT: turbinates not boggy, tympanic membranes clear, throat without erythema or postnasal drip, tonsils 1+ bilaterally, no cervical lymphadenopathy
CV: regular rate and rhythm without murmurs, rubs or gallops
Pulmonary: lungs clear to auscultation without wheezes, rhonchi, or stridor
GI: abdomen soft, non-tender
GU: no hernia appreciated, Tanner Stage 2
Neurological: no focal deficits
Psychiatric: mood normal, makes good eye contact, appears cheerful and calm
MSK: full active range of motion, normal gait, no swelling of joints

Labs:
- CBC, CMP, TSH with reflex Free T4, A1c, vitamin D levels ordered and later resulted all within normal limits.

Screening:
- NICHQ Vanderbilt Assessment Scale was taken by Mom in the office, and teacher forms were taken home. Mom was able to send the forms back. While scoring some points in inattentiveness, JR did not score high enough to be considered ADHD/ADD.

- Sleep Disturbance Scale for Children (SDSC): 20 points (positive for tiredness in the morning, difficulty waking up, and daytime somnolence but negative for other symptoms)
- Sleep Disorders Inventory for Students-Revised-Adolescent (SDIS-R-A): low risk for sleep disorders (positive for tiredness in the morning, difficulty waking up, daytime somnolence, and occasional headaches, but negative for other symptoms)

Assessment:
JR's most likely diagnosis is chronic sleep deprivation in an adolescent, who averages 7.5–8 hours a day, chronically lower than the minimum recommended for his age of 9 hours a day. While he slept more because he had less homework and activities in early elementary, now that he is in middle school, the increased cadence of his activities has affected the amount of sleep he is getting. The chronic sleep deprivation is affecting his school performance and attentiveness. While he is able to get activities done, he is taking longer for schoolwork and needs redirecting.

While he has inattentiveness and Mom was worried about the possibility of an attention deficit disorder, the issue of his inattentiveness does not appear innate, but rather it appears acquired due to lack of planning for sleep. This is further exacerbated by screentime right before bed, with blue light exposure that promotes wakefulness.

JR does not appear to have significant risk for other forms of sleep disturbance, as screening was largely unremarkable.

Plan:
- A sleep diary may be both diagnostic and therapeutic, promoting an increased consciousness of the amount of sleep he is getting, as well as helping to pre-plan sleep as a priority for his health.
- A discussion was started about picking out priority activities to pursue, perhaps playing soccer in soccer season and postponing taekwondo for the off-season. Piano seemed lower priority to JR, and a discussion was started about holding off on the instrument until Summer. Planning for naps on the weekends if needed was mentioned. Putting away screen time at least 1 hour before bed, adding an inexpensive blue light filter, changing the device lighting to "night mode" with less blue backlight was recommended. Daytime gaming could be reserved for the weekend when there was less homework. A 3-month follow-up was planned to discuss his progress.
- Mom and patient were encouraged that it was good they caught his sleep deprivation symptoms early. They were encouraged to be hopeful that with minor changes to his sleep regimen and schedule, JR could thrive again in school and enjoy more energy for the activities he enjoys.

References

1. Colten HR, Altevogt BM. Institute of Medicine (US) Committee on sleep medicine and research. Sleep disorders and sleep deprivation: an unmet public health problem. Washington, DC: National Academies Press (US); 2006.
2. Paruthi S, Brooks LJ, D'Ambrosio C, Hall WA, Kotagal S, Lloyd RM, et al. Pediatric sleep duration consensus statement: a step forward. J Clin Sleep Med. 2016;12(12):1705–6. https://doi.org/10.5664/jcsm.6368.
3. Wheaton AG, Claussen AH. Short sleep duration among infants, children, and adolescents aged 4 Months-17 years- United States, 2016-2018. MMWR Morb Mortal Wkly Rep. 2021;70:1315–21. https://doi.org/10.15585/mmwr.mm7038a1.
4. Wheaton AG, Jones SE, Cooper AC, Croft JB. Short sleep duration among middle school and high school students - United States, 2015. MMWR Morb Mortal Wkly Rep. 2018;67(3):85–90. https://doi.org/10.15585/mmwr.mm6703a1.
5. Lewin DS, Wolfson AR, Bixler EO, Carskadon MA. Duration isn't everything. Healthy sleep in children and teens: duration, individual need and timing. J Clin Sleep Med. 2016;12(11):1439–41.
6. Ohayon M, Wickwire EM, Hirshkowitz M, Albert SM, Avidan A, Daly FJ, et al. National Sleep Foundation's sleep quality recommendations: first report. Sleep Health. 2017;3(1):6–19.
7. Goel P, Goel A. Exploring the evolution of sleep patterns from infancy to adolescence. Cureus. 2024;16(7) https://doi.org/10.7759/cureus.64759.
8. Atrooz F, Salim S. Sleep deprivation, oxidative stress and inflammation. In: Donev R, editor. Advances in protein chemistry and structural biology. Academic Press; 2020. p. 309–36. https://doi.org/10.1016/bs.apcsb.2019.03.001.
9. Coles L, Thorpe K, Smith S, Hewitt B, Ruppanner L, Bayliss O, et al. Children's sleep and fathers' health and wellbeing: a systematic review. Sleep Med Rev. 2022;61:101570. https://doi.org/10.1016/j.smrv.2021.101570.
10. Elek SM, Hudson DB, Fleck MO. Couples' experiences with fatigue during the transition to parenthood. J Fam Nurs. 2002;8(3):221–40. https://doi.org/10.1177/107484070200800305.
11. Nakagawa H, Okumura N. Coordinated regulation of circadian rhythms and homeostasis by the suprachiasmatic nucleus. Proc Jpn Acad Ser B Phys Biol Sci. 2010;86(4):391–409. https://doi.org/10.2183/pjab.86.391.
12. Dijk DJ, Archer SN. Circadian and homeostatic regulation of human sleep and cognitive performance and its modulation by PERIOD3. Sleep Med Clin. 2009;4(2):111–25. https://doi.org/10.1016/j.jsmc.2009.02.001.
13. Galland BC, Taylor BJ, Elder DE, Herbison P. Normal sleep patterns in infants and children: a systematic review of observational studies. Sleep Med Rev. 2012;16(3):213–22. https://doi.org/10.1016/j.smrv.2011.06.001.
14. Chaput J-P. Nature and science of. Sleep. 2018;10:421–30.
15. Duffy JF, Willson HJ, Wang W, Czeisler CA. Healthy older adults better tolerate sleep deprivation than young adults. J Am Geriatr Soc. 2009;57(7):1245–51. https://doi.org/10.1111/j.1532-5415.2009.02303.x.
16. Matricciani L, Blunden S, Rigney G, Williams MT, Olds TS. Children's sleep needs: is there sufficient evidence to recommend optimal sleep for children? Sleep. 2013;36(4):527–34. https://doi.org/10.5665/sleep.2538.
17. Eto T, Ohashi M, Nagata K, Shin N, Motomura Y, Higuchi S. Crystalline lens transmittance spectra and pupil sizes as factors affecting light-induced melatonin suppression in children and adults. Ophthalmic Physiol Opt. 2021;41(4):900–10. https://doi.org/10.1111/opo.12809.
18. Maski KP, Kothare SV. Sleep deprivation and neurobehavioral functioning in children. Int J Psychophysiol. 2013;89(2):259–64. https://doi.org/10.1016/j.ijpsycho.2013.06.019.
19. Echevarria P, Del-Ponte B, Tovo-Rodrigues L, Matijasevich A, Halal CS, Santos IS. Screen use and sleep duration and quality at 15 years old: cohort study. Sleep Med X. 2023;5:100073. https://doi.org/10.1016/j.sleepx.2023.100073.

20. Hartstein LE, Mathew GM, Reichenberger DA, Rodriguez I, Allen N, Chang A-M, et al. The impact of screen use on sleep health across the lifespan: a National Sleep Foundation consensus statement. Sleep Health. 2024;10(4):373–84. https://doi.org/10.1016/j.sleh.2024.05.001.
21. Owens JA, Belon K, Moss P. Impact of delaying school start time on adolescent sleep, mood, and behavior. Arch Pediatr Adolesc Med. 2010;164(7):608–14. https://doi.org/10.1001/archpediatrics.2010.96.
22. Watson NF, Martin JL, Wise MS, Carden KA, Kirsch DB, Kristo DA, et al. Delaying middle school and high school start times promotes student health and performance: an American Academy of sleep medicine position statement. J Clin Sleep Med. 2017;13(04):623–5. https://doi.org/10.5664/jcsm.6558.
23. Giddens NT, Juneau P, Manza P, Wiers CE, Volkow ND. Disparities in sleep duration among American children: effects of race and ethnicity, income, age, and sex. Proc Natl Acad Sci USA. 2022;119(30):e2120009119. https://doi.org/10.1073/pnas.2120009119.
24. Schmied EA, Full KM, Lin SF, Gregorio-Pascual P, Ayala GX. Sleep health among U.S. Hispanic/Latinx children: an examination of correlates of meeting sleep duration recommendations. Sleep Health. 2022;8(6):615–9. https://doi.org/10.1016/j.sleh.2022.07.002.
25. Seton C, Fitzgerald DA. Chronic sleep deprivation in teenagers: practical ways to help. Paediatr Respir Rev. 2021;40:73–9. https://doi.org/10.1016/j.prrv.2021.05.001.
26. Conklin AI, Yao CA, Richardson CG. Chronic sleep deprivation and gender-specific risk of depression in adolescents: a prospective population-based study. BMC Public Health. 2018;18(1):724. https://doi.org/10.1186/s12889-018-5656-6.
27. American Academy of Sleep Medicine. International classification of sleep disorders. 3rd ed. Darien: American Academy of Sleep Medicine; 2014.
28. Spruyt K, Gozal D. Pediatric sleep questionnaires as diagnostic or epidemiological tools: a review of currently available instruments. Sleep Med Rev. 2011;15(1):19–32. https://doi.org/10.1016/j.smrv.2010.07.005.
29. Taylor SA. Clinical evaluation of the sleepy and sleepless patient. CONTINUUM. 2023;29(4):1031–44. https://doi.org/10.1212/con.0000000000001281.
30. Li W, Ma L, Yang G, Gan WB. REM sleep selectively prunes and maintains new synapses in development and learning. Nat Neurosci. 2017;20(3):427–37. https://doi.org/10.1038/nn.4479.
31. Bernier A, Beauchamp MH, Bouvette-Turcot AA, Carlson SM, Carrier J. Sleep and cognition in preschool years: specific links to executive functioning. Child Dev. 2013;84(5):1542–53. https://doi.org/10.1111/cdev.12063.
32. Liu J, Ji X, Pitt S, Wang G, Rovit E, Lipman T, et al. Childhood sleep: physical, cognitive, and behavioral consequences and implications. World J Pediatr. 2024;20(2):122–32. https://doi.org/10.1007/s12519-022-00647-w.
33. Simola P, Liukkonen K, Pitkäranta A, Pirinen T, Aronen ET. Psychosocial and somatic outcomes of sleep problems in children: a 4-year follow-up study. Child Care Health Dev. 2014;40(1):60–7. https://doi.org/10.1111/j.1365-2214.2012.01412.x.
34. Matthews KA, Pantesco EJ. Sleep characteristics and cardiovascular risk in children and adolescents: an enumerative review. Sleep Med. 2016;18:36–49. https://doi.org/10.1016/j.sleep.2015.06.004.
35. Vriend JL, Davidson FD, Corkum PV, Rusak B, Chambers CT, McLaughlin EN. Manipulating sleep duration alters emotional functioning and cognitive performance in children. J Pediatr Psychol. 2013;38(10):1058–69. https://doi.org/10.1093/jpepsy/jst033.
36. Sun W, Li SX, Jiang Y, Xu X, Spruyt K, Zhu Q, et al. A community-based study of sleep and cognitive development in infants and toddlers. J Clin Sleep Med. 2018;14(6):977–84. https://doi.org/10.5664/jcsm.7164.
37. Uccella S, Cordani R, Salfi F, Gorgoni M, Scarpelli S, Gemignani A, et al. Sleep deprivation and insomnia in adolescence: implications for mental health. Brain Sci. 2023;13(4):569. https://doi.org/10.3390/brainsci13040569.
38. Kocevska D, Muetzel RL, Luik AI, Luijk MP, Jaddoe VW, Verhulst FC, et al. The developmental course of sleep disturbances across childhood relates to brain morphology at age 7: the generation R study. Sleep. 2017;40(1) https://doi.org/10.1093/sleep/zsw016.

39. Skurvydas A, Zlibinaite L, Solianik R, et al. One night of sleep deprivation impairs executive function but does not affect psychomotor or motor performance. Biol Sport. 2020;37(1):7–14. https://doi.org/10.5114/biolsport.2020.89936.
40. Hershner SD, Chervin RD. Causes and consequences of sleepiness among college students. Nat Sci Sleep. 2014;6:73–84. https://doi.org/10.2147/NSS.S62907.
41. Kuula L, Pesonen AK, Martikainen S, et al. Poor sleep and neurocognitive function in early adolescence. Sleep Med. 2015;16(10):1207–12. https://doi.org/10.1016/j.sleep.2015.06.017.
42. Kamphuis J, Meerlo P, Koolhaas JM, Lancel M. Poor sleep as a potential causal factor in aggression and violence. Sleep Med. 2012;13(4):327–34. https://doi.org/10.1016/j.sleep.2011.12.006.
43. Krishnan AS, Reichenberger DA, Strayer SM, et al. Childhood sleep is prospectively associated with adolescent alcohol and marijuana use. Ann Epidemiol. 2024;98:25–31. https://doi.org/10.1016/j.annepidem.2024.07.048.
44. Touchette E, Petit D, Séguin JR, et al. Associations between sleep duration patterns and behavioral/cognitive functioning at school entry. Sleep. 2007;30(9):1213–9. https://doi.org/10.1093/sleep/30.9.1213.
45. Ramar K, Malhotra RK, Carden KA, et al. Sleep is essential to health: an American Academy of sleep medicine position statement. J Clin Sleep Med. 2021;17(10):2115–9. https://doi.org/10.5664/jcsm.9476.
46. Cappuccio FP, Taggart FM, Kandala NB, et al. Meta-analysis of short sleep duration and obesity in children and adults. Sleep. 2008;31(5):619–26. https://doi.org/10.1093/sleep/31.5.619.
47. Wu Y, Gong Q, Zou Z, Li H, Zhang X. Short sleep duration and obesity among children: a systematic review and meta-analysis of prospective studies. Obes Res Clin Pract. 2017;11(2):140–50. https://doi.org/10.1016/j.orcp.2016.05.005.
48. Miller MA, Bates S, Ji C, Cappuccio FP. Systematic review and meta-analyses of the relationship between short sleep and incidence of obesity and effectiveness of sleep interventions on weight gain in preschool children. Obes Rev. 2021;22(2):e13113. https://doi.org/10.1111/obr.13113.
49. Okoli A, Hanlon EC, Brady MJ. The relationship between sleep, obesity, and metabolic health in adolescents - a review. Curr Opin Endocr Metab Res. 2021;17:15–9. https://doi.org/10.1016/j.coemr.2020.10.007.
50. Morgan T, Basalely A, Singer P, Castellanos L, Sethna CB. The association between sleep duration and cardiometabolic risk among children and adolescents in the United States (US): a NHANES study. Child Care Health Dev. 2024;50(3):e13273. https://doi.org/10.1111/cch.13273.
51. Fatima Y, Doi SA, Mamun AA. Sleep quality and obesity in young subjects: a meta-analysis. Obes Rev. 2016;17(11):1154–66. https://doi.org/10.1111/obr.12444.
52. Baron KG, Duffecy J, Reutrakul S, et al. Behavioral interventions to extend sleep duration: a systematic review and meta-analysis. Sleep Med Rev. 2021;60:101532. https://doi.org/10.1016/j.smrv.2021.101532.
53. Irish LA, Kline CE, Gunn HE, Buysse DJ, Hall MH. The role of sleep hygiene in promoting public health: a review of empirical evidence. Sleep Med Rev. 2015;22:23–36. https://doi.org/10.1016/j.smrv.2014.10.001.
54. American Academy of Pediatrics. American Academy of Pediatrics Supports Childhood Sleep Guidelines, June 13, 2016. Retrieved on February 24, 2020.
55. Hale L, Kirschen GW, LeBourgeois MK, et al. Youth screen media habits and sleep: sleep-friendly screen behavior recommendations for clinicians, educators, and parents. Child Adolesc Psychiatr Clin N Am. 2018;27(2):229–45. https://doi.org/10.1016/j.chc.2017.11.014.
56. Ogundele MO, Yemula C. Management of sleep disorders among children and adolescents with neurodevelopmental disorders: a practical guide for clinicians. World J Clin Pediatr. 2022;11(3):239–52. https://doi.org/10.5409/wjcp.v11.i3.239.
57. Alrousan G, Hassan A, Pillai AA, Atrooz F, Salim S. Early life sleep deprivation and brain development: insights from human and animal studies. Front Neurosci. 2022;16:833786. https://doi.org/10.3389/fnins.2022.833786.

58. Qiu J, Morales-Muñoz I. Associations between sleep and mental health in adolescents: results from the UK millennium cohort study. Int J Environ Res Public Health. 2022;19(3):1868. https://doi.org/10.3390/ijerph19031868.
59. Codoñer-Franch P, Gombert M, Martínez-Raga J, Cenit MC. Circadian disruption and mental health: the chronotherapeutic potential of microbiome-based and dietary strategies. Int J Mol Sci. 2023;24(8):7579. https://doi.org/10.3390/ijms24087579.

Screen Time and Sleep

Sullafa Kadura and Poulomee Tripathi

Introduction

In today's digital age, screens have become an integral part of daily life for children and adolescents. Devices such as smartphones, tablets, televisions, or gaming consoles have increasingly found their way into the bedroom with nearly three-quarters of American children (72%) and the vast majority of adolescents (89%) have at least one electronic device in their bedroom [1]. The widespread presence of these devices is thought to adversely impact sleep through various mechanisms, including increased psychological stimulation, reduced sleep opportunities, and delayed circadian rhythms through bright light exposure [2].

Associations between screen use near bedtime—particularly with smartphones and tablets—and poorer self-reported sleep outcomes, such as inadequate sleep quantity, poor sleep quality, and increased daytime sleepiness, have been demonstrated. Additionally, simply having a television, computer, gaming console, smartphone, or tablet in the bedroom was associated with greater odds of poor self-reported sleep outcomes [2, 3]. However, many of these studies were cross-sectional in design and questionnaire-based, using self- or parent-reported measures on sleep and screen time.

Despite these limitations, these studies have informed recommendations from organizations like the American Academy of Pediatrics to limit screen use 1 to 2 hours before bed [4]. However, many adolescents continue using screens late into the night. In addition, a recent cohort study further highlighted the complex relationship between screen use and sleep, showing that while total screen use in the

S. Kadura (✉)
Department of Medicine, University of Rochester Medical Center, Rochester, NY, USA
e-mail: sullafa_kadura@urmc.rochester.edu

P. Tripathi
University of Kentucky College of Medicine, Lexington, KY, USA
e-mail: Ptr230@uky.edu

2 hours before bedtime did not reduce objective total sleep time, it was linked to delayed sleep onset and offset [5]. These findings suggest that the relationship between technology and sleep is multifaceted, involving both negative and, in some cases, potentially neutral or even positive effects.

Mechanisms of Screen Time's Impact on Sleep

The relationship between technology and sleep is more complex than our current public health messaging. Early models suggested a unidirectional negative impact, where technology use before bed was solely viewed as harmful to sleep [2]. However, recent literature has introduced a more nuanced, bidirectional relationship. In addition to negatively impacting sleep through psychological stimulation, sleep displacement, bright light exposure or night-time sleep disruption, screen use may also serve as a tool for emotional regulation or serve as a time-filler, helping some individuals cope with anxiety or pass time while waiting to feel sleepy. In practice, patients often report that screens help distract them from negative thoughts or ease worries, which can, in some cases, facilitate falling asleep [6]. See Table 2.1 for examples of mechanisms through which screens can influence sleep. Additionally, Excessive screen time has been identified as a contributing factor to childhood obesity, primarily due to increased sedentary behavior, exposure to food advertising, and disrupted eating patterns such as snacking during screen use. Children and adolescents who spend more time on screens are less likely to engage in physical activity and more likely to consume high-calorie, low-nutrient foods, leading to weight gain over time. Obesity, in turn, is a major risk factor for **sleep-disordered breathing (SDB)**, including obstructive sleep apnea. Excess weight, particularly around the neck and upper airway, can increase the risk of airway obstruction during sleep, resulting in fragmented sleep, reduced oxygen levels, and daytime fatigue. This creates a concerning cycle where screen time contributes to both poor sleep and weight gain, each of which further exacerbates the other. Addressing screen habits is therefore an important strategy not only for preventing obesity but also for reducing the risk of sleep-related breathing disorders in children.

Impact on Sleep Outcomes

Studies have shown children and adolescents who get less than the recommended duration of sleep have increased mood and behavioral issues, including working memory, depression, aggression, and decision making [7]. Therefore, it is important to understand whether screen time negatively impacts sleep.

Exposure to bright light at night leads to decreased production of melatonin. This effect has been shown to be increased in children compared to adults, likely due to pupil size and increased crystalline lens transmittance [8] and in patients with delayed sleep-wake phase circadian rhythms, where patients had a greater sustained pupil response after light exposure [9]. It should be acknowledged that not all individuals have the same sensitivity to evening/night light exposure [10]. These

Table 2.1 Mechanisms of screen time's impact on sleep

Mechanism	Description	Case Example
Psychological stimulation	Engaging screen content, like games or social media, elevates mental and physiological arousal, making it harder to eventually relax and fall asleep.	A 10-year-old plays a stimulating video game while in bed. After turning off the game, he remains mentally alert and takes longer to fall asleep than usual.
Sleep displacement	Screen time before or in bed can lead to later bedtimes and reduce total sleep duration.	An adolescent stays up playing video games, losing track of time. They go to bed later than planned but still wake up at the usual time for school, resulting in shortened sleep.
Bright Light Exposure	Evening bright light exposure from screens suppresses melatonin release, delaying sleep onset and shifting circadian timing.	A teenager uses a tablet at night, delaying sleep onset by due to melatonin suppression from bright light exposure. Although they sleep 2 hours later, they maintain enough total sleep time by waking up 2 hours later, so they have a shifted schedule.
Nighttime Sleep Disruption	Screen-related disruptions during sleep, such as notifications, that negatively impact sleep.	A teenager leaves silent mode off on their phone due to fear of missing out (FoMO) and is awakened by a notification. After checking their phone, they struggle to fall back asleep.
Time Filler	Screens are used to pass time and provide entertainment during moments of wakefulness, often when the individual isn't sleepy pre-bedtime or while in bed.	A teenager with a delayed sleep-wake phase, who naturally feels awake later in the evening, may use their phone to pass time out of boredom while lying in bed.
Emotional regulation	Screens are used to ease stress or anxiety, providing comfort and making it easier to fall asleep.	A teenager keeps the TV on for background noise to distract themselves from negative thoughts, which helps them fall asleep.

individual differences may explain why bright light is associated with overall smaller impacts on sleep onset than expected—typically between a 3- and 9-minute delay in most studies [6].

Nonetheless, there is a stronger association with subjective *pre-sleep alertness*, making individuals feel more awake before bedtime, particularly with prolonged or repeated screen use [6]. When examining objective sleep, using screens *after* getting into bed was associated with shorter sleep duration. The type of screen activity also mattered, with interactive use, such as gaming, associated with more sleep loss compared to passive use [5], highlighting the potential impact of psychological stimulation.

Although some adolescents appear to adjust their wake times to maintain a consistent total sleep duration, these phase delays in their sleep-wake cycle may be associated with an increased risk of developing sleep disorders in the future, such as delayed sleep-wake phase disorder (DSWPD) [11]. DSWPD, in turn, is associated with an increased risk of insulin resistance early in life [12].

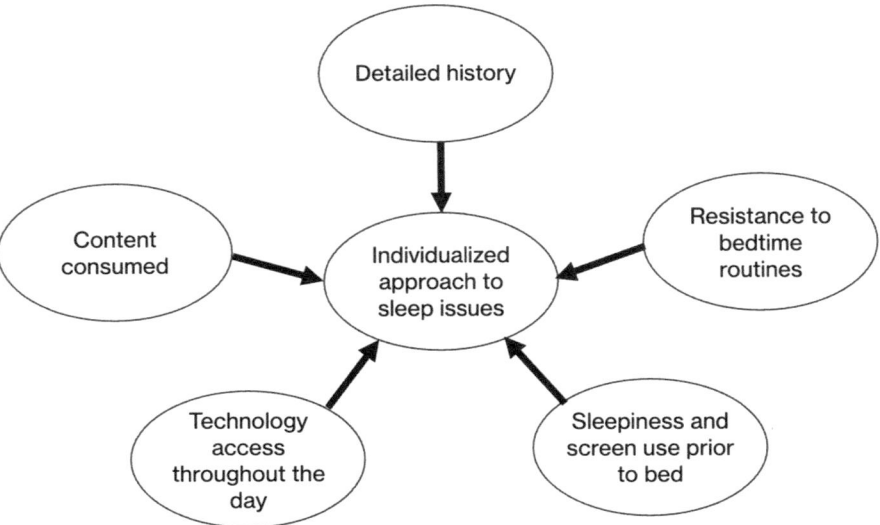

Fig. 2.1 Summary of key considerations when approaching pediatric sleep issues

Recommendations

When addressing sleep issues and screens in children and adolescents, a "one size fits all" approach does not work, and may even be harmful. Tailoring recommendations to the individual needs are crucial. Here are some key considerations (Fig. 2.1):

- *Take a detailed sleep history*:
  - Ask about their bedtime routine, what time they are going to bed, and what time they are turning the lights off to attempt sleep. Ask how long it takes to fall asleep, whether they wake up during the night, what time their final wake up is, and how they feel during the day.
  - Inquire about daytime sleepiness or attention difficulties at school.
  - These questions can help identify underlying sleep problems, such as insomnia or circadian rhythm disorders.
- *Evaluate sleepiness and screen use before bed*:
  - If it takes them a long time to fall asleep, ask if they feel sleepy when going to bed and what they are doing while trying to fall asleep.
  - If they are not sleepy and feeling bored, they may be using screens to pass the time.
  - If they are sleepy but kept awake by stress or negative thoughts, screens might be used as a tool for relaxation or distraction, which could help with sleep onset.

- *Assess resistance to bedtime routines*:
  - Is the child resistant to bedtime routines after evening screen time? This could indicate that screens are interfering with healthy sleep habits.
  - Consider recommending the removal of screen time from the evening routine in this case.
- *Ask about technology access in the bedroom*:
  - What devices are available to the child, and when are they used? Television's impact on sleep may differ from that of interactive gaming. Daytime use may not impact sleep as much as evening or night-time use.
  - Does the child wake up at night due to alerts or lighting from devices? If so, consider recommending the use of airplane mode or silent mode at night or suggest charging devices outside of the bedroom.
- *Choose content wisely*:
  - Ask if there is exposure to negative or violent content, and how the child feels about it.
  - Is the content contributing to nightmares or other sleep issues?
  - Encourage families to review video games together before purchasing and monitor how the child responds to playing them.

Recommend co-viewing media with children to help parents connect with their child and understand the impact of the content.

Conclusion

Research regarding the impact of screens on pediatric sleep has not been able to match the acceleration of technological advancement, which will continue and remain a permanent part of our modern lives. For some individuals and families, keeping devices out of the bedroom at night might be an effective strategy to minimize potential impacts on sleep. However, for others, limiting device use in this way may not be practical or easy to enforce, and in some cases, make it more difficult to fall asleep.

Clinical Pearls

- Many children and adolescents have electronics that emit bright light in their bedroom.
- Although much of the research surrounding screen time and impact on pediatric sleep is subjective/self-reported, there is correlation between screen usage and negative sleep impacts.
- Obtaining a thorough history is tantamount to ensuring whether screen usage habits, content consumed, or other pathologies are at play when children have sleep disturbances.

References

1. Gradisar M, Wolfson AR, Harvey AG, Hale L, Rosenberg R, Czeisler CA. The sleep and technology use of Americans: findings from the National Sleep Foundation's 2011 Sleep in America poll. J Clin Sleep Med. 2013;9(12):1291–9.
2. Cain N, Gradisar M. Electronic media use and sleep in school-aged children and adolescents: a review. Sleep Med. 2010;11(8):735–42.
3. Carter B, Rees P, Hale L, Bhattacharjee D, Paradkar MS. Association between portable screen-based media device access or use and sleep outcomes: a systematic review and meta-analysis. JAMA Pediatr. 2016;170(12):1202–8.
4. Guram S, Heinz P. Media use in children: American Academy of Pediatrics recommendations 2016. Arch Dis Childhood. 2018;103(2):99–101.
5. Brosnan B, Haszard JJ, Meredith-Jones KA, Wickham SR, Galland BC, Taylor RW. Screen use at bedtime and sleep duration and quality among youths. JAMA Pediatr. 2024;178:1147.
6. Bauducco S, Pillion M, Bartel K, Reynolds C, Kahn M, Gradisar M. A bidirectional model of sleep and technology use: a theoretical review of how much, for whom, and which mechanisms. Sleep Med Rev. 2024;76:101933.
7. Yang FN, Xie W, Wang Z. Effects of sleep duration on neurocognitive development in early adolescents in the USA: a propensity score matched, longitudinal, observational study. Lancet Child Adolesc Health. 2022;6(10):705–12.
8. Higuchi S, Nagafuchi Y, Lee SI, Harada T. Influence of light at night on melatonin suppression in children. J Clin Endocrinol Metab. 2014;99(9):3298–303.
9. Watson LA, Phillips AJK, Hosken IT, McGlashan EM, Anderson C, Lack LC, et al. Increased sensitivity of the circadian system to light in delayed sleep-wake phase disorder. J Physiol. 2018;596(24):6249–61.
10. Chellappa SL. Individual differences in light sensitivity affect sleep and circadian rhythms. Sleep. 2021;44(2):zsaa214.
11. Futenma K, Takaesu Y, Komada Y, Shimura A, Okajima I, Matsui K, et al. Delayed sleep-wake phase disorder and its related sleep behaviors in the young generation. Front Psych. 2023;14:1174719.
12. Koren D, O'Sullivan KL, Mokhlesi B. Metabolic and glycemic sequelae of sleep disturbances in children and adults. Curr Diab Rep. 2015;15(1):562.

Sleep, Chronotype, and Learning: A Developmental Perspective

3

Marty Martin and Zeeshawn Malik

Introduction

Primary care providers manage a broad spectrum of patient concerns—ranging from medical complaints to the growing need to dispel health-related myths and misconceptions. Sleep is one such area rife with misinformation. This chapter aims to equip primary care providers with accurate, evidence-based knowledge about sleep and practical tools to effectively communicate this information to patients and families. Emphasis is placed on delivering guidance in engaging, accessible ways—whether through face-to-face conversations or digital platforms like sleep health apps.

This chapter specifically explores the interplay between sleep, chronotype, and learning. While learning is often associated with formal education, it encompasses much more—it reflects an individual's capacity to perceive, understand, and apply information in real-world settings. Sleep plays a foundational role in supporting cognitive processes such as learning, memory, and overall performance in school, work, and daily life. These relationships are influenced by factors like biological timing, individual chronotypes (e.g., "morning larks" vs. "night owls"), developmental stage, and environmental influences such as screen use or irregular sleep routines. Understanding these dynamics within a developmental and culturally responsive framework enables primary care providers to assess, diagnose, and manage sleep-related concerns across the lifespan. It also empowers providers to offer tailored, inclusive education that supports health equity and respects the diverse backgrounds and needs of patients and their families.

This chapter is focused on the relationship among sleep, chronotype, and learning. Learning is related to education but is broader than just academic achievements.

The original version of the chapter has been revised. A correction to this chapter can be found at https://doi.org/10.1007/978-3-031-92166-7_17

M. Martin (✉)
Department of Management & Entrepreneurship, Driehaus College of Business, DePaul University, Chicago, IL, USA
e-mail: martym@depaul.edu

Z. Malik
The University of Texas at Dallas, Richardson, TX, USA

© The Author(s), under exclusive license to Springer Nature Switzerland AG 2025, corrected publication 2025
A. Wani, I. S. Khawaja (eds.), *Sleep Disorders in Children*, https://doi.org/10.1007/978-3-031-92166-7_3

It includes what an individual can perceive and understand from the environment around them and their ability to apply what they know to new environments. After reading this chapter, you understand how sleep plays a critical role these cognitive functions of learning, memory, and overall performance in school, work, and life. Sleep affects how we think, remember, and learn. This link changes based on things like what time it is, whether someone is a "morning person" or "night owl," how old they are, and outside factors such as using screens or having an irregular sleep schedule. Understanding these relationships within a developmental framework allows you as a primary care provider to diagnose, assess, and manage sleep-related issues across different age groups, providing tailored patient education that takes into account cultural diversity, competency, and health equity.

Sleep and Learning in Early Childhood: Preschool and Kindergarten

Sleep is essential for normal brain development during the formative years of life [1]. Learning and memory are part of this normal brain development [18]. During this period, children require approximately 10–13 hours of sleep per day over a 24-hour period including naps. Sleep supports synaptic plasticity, which is crucial for memory consolidation, language acquisition, and cognitive development as early as a fetus and young infant [13].

Sleep problems ranging from inadequate sleep to poor sleep quality can impair general cognitive functioning. The effects of impaired cognitive functioning can lead to difficulties in learning and behavior. Another finding to be underscored is the long-term effects of short sleep duration in early childhood on academic performance measured by grades at age 10 in elementary school. For example, it was found that children who slept on average at 2.5, 4, 5, and 6 years of age had a two to three times odds of performing below the class average in math and science even if sleep was normalized at age 10.

Memory

Memory is essential for cognitive development, learning, and functioning. There are different types of memory. The two major types are declarative and procedural memory. Declarative memory is the memorization of facts, and the empirical evidence suggests that napping among infants and young children strengthens declarative memory [19]. Procedural memory is memorization related to skills and motor activities is strengthened by napping [19].

Chronotype

Chronotype, or individual preference for morning or evening activities based upon phenotypic expression, is not as pronounced in early childhood as other developmental periods, but it begins to develop in this stage. Sleep schedules, school schedules, and extracurricular activity schedules begin to start depending upon the chronotype of the child. For instance, if a kindergartener is an evening type, emerging evidence suggests that spatial working memory (SWM) is enhanced in the late afternoon as compared to morning type kindergarteners [17].

Cultural/Family Context

Providers should consider family routines, cultural sleep practices, and environmental factors such as exposure to screens when addressing sleep issues in this age group. Research suggests that excessive screen time, especially before bed, can disrupt sleep patterns, leading to poor academic readiness [14].

The cultural and environmental context matters. Children do not live, play, and learn in a vacuum. As such, providers may also want to inquire about the chronotype of the parents/guardians. Imagine a case in which the kindergartener is an evening type, but the primary caretaker is a morning type. In this case, one of the two may be expected to be awake, alert, focused, and meet expectations at the wrong time for their genetically determined chronotype. This mismatch is called circadian misalignment.

Assessment

Primary care providers should begin sleep assessments by asking parents about their child's bedtime, wake time, night awakenings, and nap patterns. Questions should also address bedtime routines, the sleep environment, and any use of screens before bed. There are two recommended questionnaires that are psychometrically robust as it relates to validity and reliability: (1) the BEARS and CCTQ described below.

The BEARS (Bedtime, Excessive Daytime Sleepiness, Awakenings, Regularity and Duration of Sleep, Snoring) can efficiently identify potential sleep problems in a short visit [21]. The BEARS can be used for children between the ages of 2 and 12. Providers should also assess the chronotype of children using the 27-item Children's Chronotype Questionnaire [CCTQ]. The CCTQ is appropriate for children ages 4 to 11. One of the items asks the parents/guardians to select among five chronotypes: definitely a morning type, rather a morning type than evening type, neither a morning nor an evening type, rather an evening type than morning type, definitely an evening type, or I do not know. Beyond these two questionnaires, providers should also consider the family's cultural norms regarding sleep, which can influence sleep behaviors. As an example, at what age is it culturally appropriate for a child to sleep independently.

Interventions and Referrals

For mild sleep problems impacting learning and social development, providers can suggest setting regular bedtime routines, decreasing screen time before bed, and creating a calm, cool, quiet, and dark sleep environment. At this age, most of the focus is on the parents/guardians because they have control over the sleep environment, bedtime routines, and also wake up/go to bedtimes. If these interventions are ineffective, then refer to a pediatric sleep specialist or a child psychologist with expertise in sleep.

Elementary School: Ages 6–12

As children enter elementary school, children require approximately 9–12 hours of sleep per day over a 24-hour period including naps. Sleep remains critical for attention, executive function, and academic performance [7]. Studies have shown that children with adequate sleep perform better in school, exhibiting improved attention, memory, and problem-solving skills [3]. Declarative memory is strengthened after sleeping overnight [19]. Cultural considerations, such as bedtime routines and the role of family, should be taken into account when managing sleep issues in this age cohort.

Assessment

During well-child visits, providers should routinely ask about the child's sleep duration, bedtime and wake time consistency, and any difficulties falling or staying asleep. Brief assessments can include questions about the child's daytime functioning, such as attention and mood, which can be impacted by poor sleep. Sleep diaries completed by parents for a week can be useful tools to identify patterns. The BEARS can be used for children between the ages of 2 and 12. The CCTQ is appropriate for children ages 4–11.

Inquiries about school behavior by the primary provider or another primary care clinical staff should focus upon attendance, tardiness, dismissal, detention, and academic performance. Additionally, inquiries about any changes in the sleep environment (i.e., sleeping alone or with parents/caretakers), as well as study environment at home should also be explored. Parents/caretakers should be asked about the length of the school day from wake-up time, commuting to school, means of commuting to coming home after school, and/or an after-school program.

Interventions and Referrals

Recommendations should include reinforcing consistent sleep schedules, particularly maintaining regular bedtimes and wake times even on weekends. Providers can advise limiting caffeine and screen exposure in the evening. For children experiencing more persistent issues, such as insomnia or sleep apnea, referral to a pediatric sleep specialist or a behavioral sleep medicine program may be warranted. Considerations for telehealth can include virtual guidance on sleep hygiene and behavioral techniques.

Middle School: Ages 12–14

In middle school, sleep needs remain at about 8–10 hours per night. In one study, it was found that 8 hours of sleep was associated with the highest levels of academic performance based on math grades. In short, there is a curvilinear relationship between sleep duration and academic outcomes. However, sleep duration often declines due to academic pressures, extracurricular activities, and increased use of technology, especially social media and gaming, which can delay bedtimes and reduce sleep quality. Attention and memory are empirically associated with school performance among middle-school students.

Chronotype

The circadian rhythm also shifts later during puberty, leading to a preference for eveningness. This shift, coupled with early school start times, often results in chronic sleep deprivation and social jet lag, affecting learning, memory, and mood. Social jet lag is the mismatch between your genetically determined chronotype and the demands of the outside world including going to school. Generally speaking, the greater the difference in wake-up time during school/work days and non-school/work days, then the greater negative effects of social jet lag. Ideally, this difference should be no more than 1 to 1.5 hours.

Cultural/Family Context

Culturally, there may be variations in attitudes toward bedtime and sleep hygiene, which can influence adherence to sleep recommendations. Providers should explore these cultural contexts when discussing sleep with families, as culturally tailored interventions may be more effective in promoting healthy sleep habits. Family members and also media figures are modeling appropriate sleep behaviors, sleep timing and the connection or lack of connection between enough, high quality sleep, and academic performance.

Assessment

Providers should assess sleep in adolescents by inquiring about bedtimes, wake times, screen use, and overall sleep duration. Asking about school start times, daytime sleepiness, academic performance, and mood can also provide insights into the impact of sleep on daily functioning. The use of tools such as the Adolescent Sleep Wake Scale (ASWS) can provide a more detailed understanding of sleep quality in this age group. Another assessment is the 8-item Pediatric Daytime Sleepiness Scale appropriate for ages 11–15 [9].

Interventions and Referrals

For minor sleep issues, interventions can include education on the impact of screens and social media on sleep and encouraging the establishment of a bedtime routine free of electronic devices. Providers should emphasize the importance of sleep for academic performance and mood regulation, offering resources tailored to adolescents. If significant problems like insomnia, delayed sleep phase syndrome, or significant social jet lag are identified, referrals to a pediatric sleep specialist or a behavioral health provider with expertise in adolescent sleep may be necessary. For telehealth visits, virtual sleep coaching or behavioral interventions can be beneficial.

High School: Ages 14–18

High-school students require 8–10 hours of sleep, but most adolescents do not achieve this due to late bedtimes and early school start times. This age group is particularly affected by social jet lag, which exacerbates the misalignment between their biological clocks and societal demands. Insufficient sleep and irregular sleep are commonplace among this age cohort [8].

Insufficient sleep in adolescents age has been linked to poorer academic performance, reduced cognitive function, and increased risk of mood disorders. Declarative memory is strengthened both immediately after leaning and sleeping overnight among high-school students [19].

Chronotype

Chronotype begins to shift during this period, with some children showing a preference for eveningness, which can conflict with early school start times. This misalignment can result in social jet lag, where children's biological clocks are out of sync with their school schedules, potentially impacting learning and behavior.

Technology

Technology use significantly impacts sleep in this age group. The blue light emitted from screens can suppress melatonin production, delaying sleep onset and reducing sleep quality [6]. Additionally, cultural/friendship norms around technology use, such as constant connectivity and social media engagement, can further disrupt sleep patterns. Health equity considerations are important, as students from lower socioeconomic backgrounds may face additional barriers to achieving adequate sleep, such as environmental stressors or limited access to quiet, safe sleep environments [10].

It is recommended that school start times for high-school students begin no earlier than 8:30 am given the empirical evidence that later school start times are associated with longer sleep and less daytime sleepiness [20].

Assessment

High-school students often present with sleep issues related to social jet lag, late bedtimes, and irregular sleep schedules. Providers should ask about sleep habits, screen time, caffeine use, and any school-related stressors that might affect sleep. The Sleep Disturbance Scale for Children (SDSC) can be adapted for use with older adolescents to identify specific sleep disorders [4].

Interventions and Referrals

Interventions should focus on improving sleep hygiene, such as establishing consistent sleep-wake times, reducing evening screen exposure, and limiting caffeine. Providers can educate teens on the risks of sleep deprivation, including its impact on learning and mental health. If more complex sleep issues such as insomnia or significant circadian rhythm disorders are suspected, referral to a sleep specialist is recommended. Telehealth can be utilized for follow-up on behavioral interventions or for initial assessments that do not require in-person visits.

College/University: Ages 18–22

College students continue to experience shifts toward eveningness, with many reporting inconsistent sleep schedules due to academic demands, social activities, and part-time work [15]. The recommended sleep duration for this age group is 7–9 hours, but many students fall short, often compensating with irregular sleep patterns that impair learning and memory consolidation [11].

Chronotype

The effects of social jet lag and technology use are pronounced in college students, with many engaging in night-time use of devices for both academic and social purposes. This population is also at risk for sleep disorders, such as insomnia and delayed sleep phase disorder, which can further compromise academic performance [2].

Cultural/Family Context

Culturally responsive approaches that address the diverse backgrounds and stressors of college students can help in managing these sleep-related issues. In this age cohort, formal learning may take place in community college, university, the military, job corps training, trade school, apprentice program or "on the job" in a variety of industries, and a wide array of roles. As such, primary care providers need to recognize that in this age group the living, studying, learning, and working context can change quite dramatically and become a source of disturbed sleep accounting for changes in learning, memory, and academic/work/apprentice/recruit performance.

Assessment

Providers should assess sleep patterns in college students by discussing bedtimes, wake times, variability in sleep schedules, and the impact of academic and social commitments on sleep. Brief sleep questionnaires, such as the Pittsburgh Sleep Quality Index (PSQI), can help identify poor sleep quality [5].

Interventions and Referrals

Direct interventions include educating students on the importance of regular sleep schedules, even on weekends, and the impact of technology on sleep. Providers can offer strategies for managing sleep in the context of academic pressures, such as time management and relaxation techniques. If significant sleep disorders like insomnia or delayed sleep phase disorder are identified, referrals to sleep specialists or behavioral sleep programs are advisable. Telehealth services can be a practical way to provide ongoing support and interventions.

Adults: Ages 21–65

Adults generally require 7–9 hours of sleep per night, although individual needs can vary [16]. Sleep quality and duration in adults affect cognitive functions, including learning and memory.

Chronotype

Chronotype differences persist, with evening types often experiencing difficulties aligning their schedules with societal norms, leading to social jet lag and associated cognitive impairments.

Technology use continues to impact sleep in adults, with the proliferation of smart devices and the demands of a 24/7 connected society. Cultural factors, such as work schedules, family responsibilities, and societal expectations play a significant role in shaping sleep patterns. Health equity considerations are crucial, as adults from disadvantaged backgrounds may experience greater sleep disturbances due to stress, shift work, or limited access to healthcare resources [12].

Assessment

In adults, sleep assessments should include questions about sleep duration, quality, regularity, and any symptoms of sleep disorders such as insomnia or obstructive sleep apnea. The Epworth Sleepiness Scale (ESS) can quickly screen for excessive daytime sleepiness. In telehealth settings, virtual questionnaires can be administered before or during the appointment to maximize time efficiency.

Interventions and Referrals

Interventions for adults may involve sleep hygiene education, cognitive behavioral therapy for insomnia (CBT-I), and guidance on managing technology use before bedtime. For adults with suspected sleep apnea, referrals for sleep studies or to a sleep specialist are necessary. Providers should also consider cultural and socioeconomic factors that may affect access to sleep resources, offering referrals to community-based programs or online resources as appropriate.

Conclusion

Sleep is a fundamental component of learning, memory, and academic performance across all stages of development. Understanding the interplay between sleep, chronotype, and external influences such as technology and social responsibilities can help primary care providers develop effective strategies for managing sleep-related issues in their patients. By considering cultural diversity and health equity, providers can offer more personalized and effective care, ultimately supporting better cognitive, social, and academic outcomes for individuals across all ages.

References

1. Alrousan G, Hassan A, Pillai AA, Atrooz F, Salim S. Early life sleep deprivation and brain development: insights from human and animal studies. Front Neurosci. 2022;16:833786.
2. Becker SP, Sidol CA, Van Dyk TR, Epstein JN, Beebe DW. Intraindividual variability of sleep/wake patterns in relation to child and adolescent functioning: a systematic review. Sleep Med Rev. 2017;34:94–121.

3. Bowers JM, Moyer A. Effects of school start time on students' sleep duration, daytime sleepiness, and attendance: a meta-analysis. Sleep Health. 2017;3(6):423–31.
4. Bruni O, Ottaviano S, Guidetti V, Romoli M, Innocenzi M, Cortesi F, Giannotti F. The sleep disturbance scale for children (SDSC) construct ion and validation of an instrument to evaluate sleep disturbances in childhood and adolescence. J Sleep Res. 1996;5(4):251–61.
5. Buysse DJ, Reynolds CF III, Monk TH, Berman SR, Kupfer DJ. The Pittsburgh Sleep Quality Index: a new instrument for psychiatric practice and research. Psychiatry Res. 1989;28(2):193–213.
6. Chang AM, Aeschbach D, Duffy JF, Czeisler CA. Evening use of light-emitting eReaders negatively affects sleep, circadian timing, and next-morning alertness. Proc Natl Acad Sci. 2015;112(4):1232–7.
7. Dewald JF, Meijer AM, Oort FJ, Kerkhof GA, Bögels SM. The influence of sleep quality, sleep duration and sleepiness on school performance in children and adolescents: a meta-analytic review. Sleep Med Rev. 2010;14(3):179–89.
8. Diogo FMC, Bessa ZCM, Galina SD, de Oliveira MLC, da Silva-Júnior ELR, Valdez P, de Azevedo CVM. Sex differences in temporal sleep patterns, social jetlag, and attention in high school adolescents. Sleep Sci. 2024;17(02):e125–33.
9. Drake C, Nickel C, Burduvali E, Roth T, Jefferson C, Badia P. The pediatric daytime sleepiness scale (PDSS): sleep habits and school outcomes in middle-school children. Sleep. 2003;26(4):455–8.
10. El-Sheikh M, Kelly RJ, Buckhalt JA, Benjamin Hinnant J. Children's sleep and adjustment over time: the role of socioeconomic context. Child Dev. 2010;81(3):870–83.
11. Gaultney JF. The prevalence of sleep disorders in college students: impact on academic performance. J Am Coll Heal. 2010;59(2):91–7.
12. Grandner MA, Patel NP, Gehrman PR, Perlis ML, Pack AI. Problems associated with short sleep: bridging the gap between laboratory and epidemiological studies. Sleep Med Rev. 2010;14(4):239–47.
13. Graven SN, Browne JV. Sleep and brain development: the critical role of sleep in fetal and early neonatal brain development. Newborn Infant Nurs Rev. 2008;8(4):173–9.
14. Hale L, Guan S. Screen time and sleep among school-aged children and adolescents: a systematic literature review. Sleep Med Rev. 2015;21:50–8.
15. Hershner SD, Chervin RD. Causes and consequences of sleepiness among college students. Nat Sci Sleep. 2014;6:73–84.
16. Hirshkowitz M, Whiton K, Albert SM, Alessi C, Bruni O, DonCarlos L, et al. National Sleep Foundation's sleep time duration recommendations: methodology and results summary. Sleep Health. 2015;1(1):40–3.
17. Jafar A, Nur K, et al. Chronotype and time-of-day effects on spatial working memory in preschool children. J Clin Sleep Med. 2023;19(10):1717–26.
18. Jiang F. Sleep and early brain development. Ann Nutr Metab. 2019;75(Suppl. 1):44–54.
19. Mason GM, Lokhandwala S, Riggins T, Spencer RM. Sleep and human cognitive development. Sleep Med Rev. 2021;57:101472.
20. Meltzer LJ, Wahlstrom KL, Plog AE, Strand MJ. Changing school start times: impact on sleep in primary and secondary school students. Sleep. 2021;44(7):zsab048.
21. Owens JA, Dalzell V. Use of the 'BEARS'sleep screening tool in a pediatric residents' continuity clinic: a pilot study. Sleep medicine. 2005;6(1):63–9.

School Start Times

4

Tanya Martinez-Fernandez, Rachel Manuel, Zaiba Jetpuri, and Swetha Gannarapu

Introduction

School start times are a powerful yet often overlooked lever in promoting pediatric sleep health. Decades of research show that early school start times are biologically misaligned with the natural sleep-wake rhythms of adolescents, contributing to chronic sleep deprivation and its associated cognitive, emotional, and physical health consequences. As a primary care provider, you are uniquely positioned to advocate for change—not only within the exam room but also within your community. Your voice carries weight. By educating families, engaging with school boards, and collaborating with local stakeholders, you can play a pivotal role in driving policy change toward healthier school schedules. This chapter explores the science behind adolescent sleep patterns, the impact of early start times, and practical strategies for advocacy. Equipped with evidence and a call to action, you have the potential to shift systems and protect the sleep—and futures—of the children you care for.

The pressure associated with school and social schedules can frequently lead to suboptimal sleep in young children and adolescents. Elementary school children with good sleep habits usually sleep longer than their middle school and high-school counterparts [1]. Therefore, younger children routinely fall asleep and wake up earlier. Since most primary public schools in the United States start before 8:30 am, a healthy sleep schedule for elementary school children would include a bedtime of 7–8 pm [2]. As children reach adolescence, they can stay awake longer, and the

T. Martinez-Fernandez (✉) · S. Gannarapu
Department of Pediatrics, Pediatric Pulmonology and Sleep Medicine, UT Southwestern Medical Center, Dallas, TX, USA
e-mail: tanya.martinez@utsouthwestern.edu; Swetha.Gannarapu@utsouthwestern.edu

R. Manuel · Z. Jetpuri
Department of Family and Community Medicine, UT Southwestern Medical Center, Dallas, TX, USA
e-mail: rachel.manuel@utsouthwestern.edu; zaiba.jetpuri@utsouthwestern.edu

average natural bedtime becomes closer to 11 pm. School start times, however, can result in earlier than desired wake times and insufficient sleep in older children and teenagers.

National surveys and polls on sleep in children provide insight into typical sleep patterns of American youth. According to the National Survey of Children's Health conducted from 2016 to 2018 [3], 35% of parents reported that their children between 4 months and 17 years of age slept for less time than recommended (Fig. 4.1). Among school-aged children, adolescents (aged 13 to 17 years) displayed the highest prevalence of parent-reported shortened sleep duration at 31.2%. The 2006 Sleep in America poll [3], conducted by the National Sleep Foundation, found that sixth graders tended to sleep for an average of 8.4 hours, while 12th graders only averaged 6.9 hours [4]. Over the past several decades, adolescent sleep challenges have persisted despite attempts to address and educate the public on the physical and mental benefits of healthy sleep. In a recent survey, 80% of teens noted not getting enough sleep [5]. Additionally, teens who have difficulties sleeping reported more frequent depressive symptoms compared to those exercising healthy sleep habits and sleep duration [5].

The deleterious effects of inadequate sleep become more apparent as children enter adolescence when puberty-related biological changes result in a circadian rhythm phase delay, and thus, a natural shift to a later sleep onset time [6]. Chronotype is an individual's natural preference of circadian typology and is divided into "morningness" or "eveningness" with teenagers having a biological preference for delayed phase or "eveningness". Studies have demonstrated that dim light melatonin onset (DLMO), a marker for a person's biological night, positively correlates

Fig. 4.1 Healthy sleep duration infographic from the American Academy of Sleep Medicine. Recommendations for total hours of sleep among age groups. Minimum (green bars) to maximum (orange bars) daily total hours of sleep represented including naps for children 4 m-5 years of age (*) [31]

with tanner stage [7] and later DLMO is associated with advanced tanner staging. Several hypotheses explain the inherent phase delay that occurs during puberty, including increased sensitivity to light [8], increased internal day length [9], delay in endogenous release of melatonin, and a slower build of sleep drive in adolescence [8]. Despite natural circadian shifts, adolescents do not have a reduced need for sleep [7], with 9 hours being the average recommended amount of sleep for teens.

Regular and consistent sleep duration is critical in maintaining physical and mental health. Obesity trends in the United States show increased prevalence in all pediatric age groups (Fig. 4.2). However, obesity prevalence is significantly higher in adolescents, Hispanic, and non-Hispanic black children, foreboding a tendency to population health disparities [10]. Stable circadian rhythm patterns are essential in equilibrating stress responses that affect growth and development through the hypothalamic-pituitary-adrenal (HPA) axis. Sleep disruption promotes an imbalance in HPA axis hormone release in cortisol and growth hormone, which are, in turn, regulated by other hormones, including leptin and ghrelin. Dysregulation of leptin and ghrelin, hormones that regulate hunger and satiety, associated with reduced sleep duration, exacerbates metabolic derangements that result in poor health outcomes such as obesity, type 2 diabetes, and hypertension. Furthermore, abnormal fluctuations of leptin and ghrelin result in overeating and insulin resistance [11, 12]. Children who sleep less are thus more likely to have increased caloric intake of non-core foods, such as sweets and snacks, resulting in higher BMI [13].

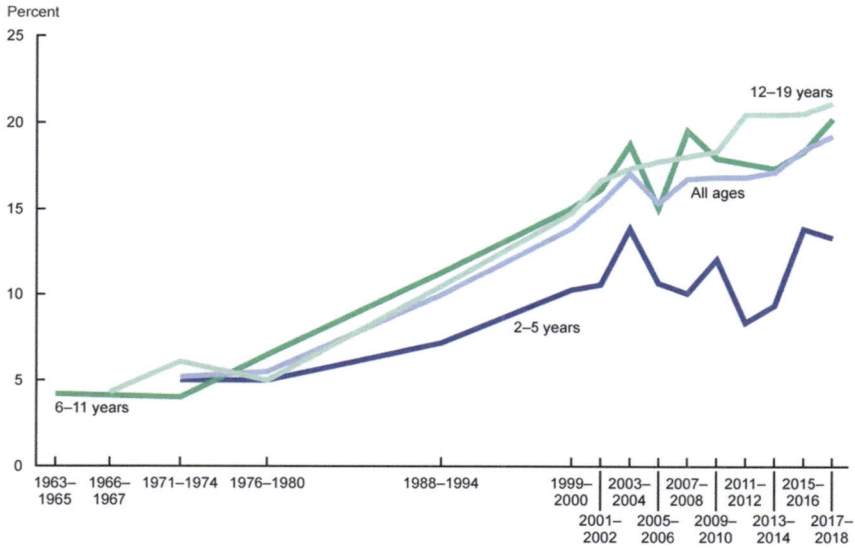

Fig. 4.2 Pediatric obesity trends in children and adolescents. (Source: National Center for Health Statistics)

An association between alterations in the immune response and higher susceptibility to infection in sleep-deprived subjects further highlights the importance of healthy sleep duration [14]. Animal models have demonstrated sleep-derived alterations in immunity. Translocation of pathogenic bacteria resulting in bacteremia and lymphatic invasion, in a group of sleep-deprived rats, was hypothesized to result from immune dysregulation associated with abnormal sleep duration [15]. Inflammatory cytokines, including interleukin-1 (IL-1) and tumor necrosis factor-alpha (TNF alpha), appear to regulate sleep, while certain microbe-related factors activate these factors or regulate sleep themselves [14]. In an observational study by Orzech and co-workers [16], adolescents with short sleep duration were found to be more susceptible to infections including gastroenteritis, cold, and flu symptoms than those with longer sleep duration.

Research implicates not only physical but also emotional well-being to insufficient sleep. Eight out of 10 teens do not get enough sleep, and 75% of teens report their well-being being negatively impacted when they sleep less than the recommended standards [5]. Symptoms of depression and anxiety are more evident in those who get less than 6 hours of sleep a night. Mental health effects impact learning through poor memory consolidation, decreased attention span, and reduced ability to problem solve, including in those who are high achievers. These detrimental cognitive effects are particularly concerning in school-aged children as they contribute to declining academic performance. In addition to impaired cognitive function, adverse behaviors can occur including difficulty with impulse control, slow response times, increased irritability, and frustration intolerance which have broader psychosocial implications. In teenagers, particularly, adverse effects on mood and behavior can have physical safety associations, such as an increase in car accidents, fighting, and substance use [17].

School athletes are particularly vulnerable to sleep deprivation, with sleep duration being the strongest predictor of adolescent sports injury. Students sleeping less than 8 hours are 70% more likely to be injured [18]. NREM sleep results in physiologic recuperation by generating proteins, muscle growth, and repair. Lack of sleep can result in prolonged reaction times, reduced attention span, impaired fine motor coordination, and impaired decision-making. Athletes with reduced sleep time report higher fatigue levels than those who sleep longer [19]. Despite evidence that longer sleep improves athletic performance, elite college student-athletes identify that athletic commitments commonly interfere with sleep quality and quantity [20]. More recently, the severity of depression in athletes has been associated with poor sleep quality and difficulty maintaining sleep adding to evidence that depressive symptoms can be amplified in those with insufficient sleep [21]. Student athletes have challenging school and practice schedules making them a unique population when considering sleep healthy school policy.

Public Health Policy for Later School Start Times

Extensive evidence demonstrates the benefits of later school start times to middle and high-school children. In fact, since the initial studies by Carskadon and co-workers [5–8] on sleep phase delay during adolescence, multiple organizations including the National Sleep Foundation, the American Academy of Sleep Medicine, and the American Academy of Pediatrics officially recommend a school start time of no earlier than 8:30 am for middle and high schools. Despite these efforts, the average school start time remains at 8:00 am across public high schools in the United States, with the earliest being 7:30 am [2]. In a recent poll among U.S. teens, 70% reported having to be in school before 8:30 am. Delays in school start times among high-school adolescents mitigate the consequences of sleep deprivation (Table 4.1) including reduced daytime sleepiness and tardiness [22].

Following Carskadon's pivotal work in pubertal sleep-related changes, public health officials began recognizing that later high school start times could be a modifiable health risk factor. In a pioneer public health policy recommendation, the former director of the Minnesota Medical Association encouraged high schools to delay start times for teenagers. Several high-school districts in the state adopted delayed start times from 7:15–7:20 to 8:30–8:40 am. A review following the policy change demonstrated an increase in attendance, improvement in grades, and nearly an hour of extra sleep by teens in districts with a move to the later start times compared to those that did not [23]. Investigators have described improvements in behavior and academic achievement as well as reduction in substance abuse and mental health [23–26].

Middle schoolers similarly benefit from delayed start times [27]. Although higher degree of variability exists in pubertal changes and sex differences in this group, current data in middle-school subjects also supports a school start time later than 8:30 am. An analysis of U.S. and international studies in middle-school subjects concluded that children with later than usual school start times demonstrated improved sleep health, behavior, and academic performance. These older children were more likely to sleep 8 hours, less likely to be tardy or absent, and were less sleepy during the day. Academically, students had better cognitive skills including

Table 4.1 Benefits of delayed start school times
Improved total sleep time
Less daytime sleepiness
Less tardiness
Reduction in childhood obesity
Safe driving
Faster reaction times on performance testing
Improved mood
Higher grades and average GPA
Lower average depressive/anxiety symptoms

improved GPA and standardized test scores. Affect was more positive in those with a longer sleep opportunity with reduced feelings of sadness and hostility.

Benefits incurred from delayed school start times may extend beyond the broader population, including teachers and parents. In Colorado, high-school teachers reported long-term improvements in sleep quality and duration after implementing later school start times. Perceived good sleep quality increased to 48% of respondents after implementation compared to 29%. In addition, teachers were less likely to feel tired during the day (49% pre- to 29% post-intervention) [28]. Parental effects were found to be more variable depending on what grades their children were in. Most benefit was seen in parents of high-school students who report improved sleep duration, good sleep quality, and less daytime tiredness than those with elementary school children. Parents with multiple children in different grades did not have appreciable changes to their sleep cycle [29].

Adoption of healthy sleep public policy will require work addressing economic and logistical concerns from local and state governments, school districts, and parents. Arguments against later school start times have centered around concern for economic costs to school districts and parents. Resolving issues such as coordinating school transportation to allow for different pick-up times may cost school districts more. However, these costs may be offset by long-term gains. An economic analysis sponsored by RAND, a nonprofit, non-partisan research organization, concluded that long-term economic benefits outweighed immediate costs. They estimated that moving school start times would result in improved academic performance and reduced motor vehicle accidents with an influx of $83 billion to the US economy over a decade [30]. In addition, further studies are needed in non-urban populations and historically underrepresented populations, such as racial and ethnic minoritized groups, who may benefit most health-wise and economically from changes in public policy but have mitigated voices during decision making.

Call for Action

Compelling evidence exists that insufficient sleep impacts health deleteriously and biologic changes during puberty affect circadian rhythms providing the basis for public health and education policy. In 2019, California became the first state to enact legislation delaying school start times across the state. A minority of other states have since passed or are considering similar legislation. Although public policy has shifted, as of 2018, less than 20% of public high schools in the United States had start times later than or equal to 8:30 [2].

Take-Home Points

To better support healthy sleep, an age-based approach to school start times results in improved academic, physical, and mental health outcomes. The circadian phase delay associated to puberty places middle and high-school students at high risk for poor sleep duration due to social requirements and constraints. Average school start times in the United States leave teens vulnerable to physical and mental effects of sleep deprivation including metabolic dysregulation and depression. Fortunately, public policy leading to modest delays in school start times to 8:30 am for older children and teens results in population health benefits. Despite extensive data supporting later school start times in middle and high school, a large proportion of school districts have yet to adopt later school start times. Ongoing efforts are needed to expand adoption of later school start times for middle and high-school students in the United States.

References

1. Wheaton AG, et al. Short sleep duration among infants, children, and adolescents aged 4 months-17 years - United States, 2016–2018. MMWR Morb Mortal Wkly Rep. 2021;70(38):1315–21.
2. Online Archive of National Center for Education Statistics (Internet) - (cited 2024 Dec). Available from: https://nces.ed.gov/pubs2020/2020006/index.asp.
3. Online Archive of National Survey of Children's Health (Internet) - (cited 2024 Dec). Available from: https://www.childhealthdata.org/learn-about-the-nsch/NSCH.
4. Online Archive National Sleep Foundation. Sleep in America Poll: Teens and Sleep (Internet). Washington, DC: National Sleep Foundation; 2006 (cited 2024 Dec). Available from: https://www.thensf.org/wp-content/uploads/2021/03/2006-SIA-summary_of_findings.pdf.
5. Online Archive National Sleep Foundation. Sleep in America Poll: Teens' sleep Health and Mental Health are Strongly Connected (Internet). Washington, DC: National Sleep Foundation; 2024—(cited 2024 Dec). Available from: https://www.thensf.org/sleep-in-america-polls/.
6. Carskadon MA, Vieira C, Acebo C. Association between puberty and delayed phase preference. Sleep. 1993;16(3):258–62. https://doi.org/10.1093/sleep/16.3.258.
7. Carskadon MA. Sleep in adolescents: the perfect storm. Pediatr Clin North Am. 2011;58(3):637–47. https://doi.org/10.1016/j.pcl.2011.03.003.
8. Hagenauer MH, Perryman JI, Lee TM, Carskadon MA. Adolescent changes in the homeostatic and circadian regulation of sleep. Dev Neurosci. 2009;31(4):276–84. https://doi.org/10.1159/000216538.
9. Carskadon MA, Labyak SE, Acebo C, Seifer R. Intrinsic circadian period of adolescent humans measured in conditions of forced desynchrony. Neurosci Lett. 1999;260(2):129–32. https://doi.org/10.1016/s0304-3940(98)00971-9.
10. Stierman B, Afful J, Carroll MD, et al. National Health and Nutrition Examination Survey 2017–March 2020 prepandemic data files development of files and prevalence estimates for selected health outcomes. Natl Health Stat Report. 2021;158.
11. Schmid SM, Hallschmid M, Jauch-Chara K, Born J, Schultes B. A single night of sleep deprivation increases ghrelin levels and feelings of hunger in normal-weight healthy men. J Sleep Res. 2008;17(3):331–334. https://doi.org/10.1111/j.1365-2869.2008.00662.x. Epub 2008 Jun 28.
12. Mosavat M, Mirsanjari M, Arabiat D, Smyth A, Whitehead L. The role of sleep curtailment on leptin levels in obesity and diabetes mellitus. Obes Facts. 2021;14(2):214–21. https://doi.org/10.1159/000514095. Epub 2021 Mar 23. PMID: 33756469; PMCID: PMC8138234.
13. Haszard JJ, Jackson R, Morrison S, et al. Losing sleep influences dietary intake in children: a longitudinal compositional analysis of a randomised crossover trial. Int J Behav Nutr Phys Act. 2024;21:61. https://doi.org/10.1186/s12966-024-01607-5.
14. Garbarino S, Lanteri P, Bragazzi NL, et al. Role of sleep deprivation in immune-related disease risk and outcomes. Commun Biol. 2021;4:1304. https://doi.org/10.1038/s42003-021-02825-4.

15. Everson CA, Toth LA. Systemic bacterial invasion induced by sleep deprivation. Am J Physiol Regul. 2000;278:R905–16.
16. Orzech KM, Acebo C, Seifer R, Barker D, Carskadon MA. Sleep patterns are associated with common illness in adolescents. J Sleep Res. 2014;23(2):133–42. https://doi.org/10.1111/jsr.12096. Epub 2013 Oct 18. PMID: 24134661; PMCID: PMC4115328.
17. Wong MM, Robertson GC, Dyson RB. Prospective relationship between poor sleep and substance-related problems in a national sample of adolescents. Alcohol Clin Exp Res. 2015;39(2):355–62. https://doi.org/10.1111/acer.12618. Epub 2015 Jan 16. PMID: 25598438; PMCID: PMC4331208.
18. Milewski MD, Skaggs DL, Bishop GA, Pace JL, Ibrahim DA, Wren TA, Barzdukas A. Chronic lack of sleep is associated with increased sports injuries in adolescent athletes. J Pediatr Orthop. 2014;34(2):129–33. https://doi.org/10.1097/BPO.0000000000000151.
19. Sargent C, Lastella M, Halson SL, Roach GD. The impact of training schedules on the sleep and fatigue of elite athletes. Chronobiol Int. 2014;31(10):1160–8. https://doi.org/10.3109/07420528.2014.957306.
20. Penn Schoen Berland. Student-Athlete Time Demands. 2015.
21. Kim K, Huskey A, Taylor D. 0641 Difficulty maintaining sleep and sleep quality mediate the relationship between sleep disturbances and depression. Sleep. 2023. 46(Supplement_1):A282. https://doi.org/10.1093/sleep/zsad077.0641.
22. Bowers JM, Moyer A. Effects of school start time on students' sleep duration, daytime sleepiness, and attendance: a meta-analysis. Sleep Health. 2017;3(6):423–31. https://doi.org/10.1016/j.sleh.2017.08.004. Epub 2017 Sep 28.
23. Wahlstrom KL, Berger AT, Widome R. Relationships between school start time, sleep duration, and adolescent behaviors. Sleep Health. 2017;3(3):216–21. https://doi.org/10.1016/j.sleh.2017.03.002. Epub 2017 Apr 8. PMID: 28526260; PMCID: PMC7178613.
24. Owens JA, Belon K, Moss P. Impact of delaying school start time on adolescent sleep, mood, and behavior. Arch Pediatr Adolesc Med. 2010;164(7):608–14. https://doi.org/10.1001/archpediatrics.2010.96.23.
25. Wahlstrom K, Dretzke B, Gordon M, Peterson K, Edwards K, Gdula J. Examining the impact of later high school start times on the health and academic performance of high school students: A multi-site study. http://conservancy.umn.edu/handle/11299/162769; 2014.
26. Pasch KE, Laska MN, Lytle LA, Moe SG. Adolescent sleep, risk behaviors, and depressive symptoms: are they linked? Am J Health Behav. 2010;34(2):237–48.
27. Barlaan DR, Pangelinan BA, Johns A, Schweikhard A, Cromer LD. Middle school start times and young adolescent sleep, behavioral health, and academic performance outcomes: a narrative review. J Clin Sleep Med. 2022;18(11):2681–94. https://doi.org/10.5664/jcsm.10224. PMID: 35946417; PMCID: PMC9622981.
28. Wahlstrom KL, Plog AE, McNally J, Meltzer LJ. Impact of changing school start times on teacher sleep health and daytime functioning. J Sch Health. 2023;93(2):128–34. https://doi.org/10.1111/josh.13254. Epub 2022 Nov 6.
29. Meltzer LJ, Wahlstrom KL, Plog AE, Strand MJ. Changing school start times: impact on sleep in primary and secondary school students. Sleep. 2021;44(7):zsab048. https://doi.org/10.1093/sleep/zsab048.. PMID: 33855446; PMCID: PMC8271142.
30. Online Archive of RAND (Internet) - (cited 2024 Dec). Available from: https://www.rand.org/randeurope/research/projects/2017/economic-benefits-of-later-school-start-times.html
31. Paruthi S, Brooks LJ, D'Ambrosio C, Hall WA, Kotagal S, Lloyd RM, Malow BA, Maski K, Nichols C, Quan SF, Rosen CL, Troester MM, Wise MS. Recommended amount of sleep for pediatric populations: a consensus statement of the American Academy of Sleep Medicine. J Clin Sleep Med. 2016;12(6):785–6. https://doi.org/10.5664/jcsm.5866. PMID: 27250809; PMCID: PMC4877308.

Part II
Sleep Through the Years

Infant Sleep

Likhita Shaik, Daniel Rongo, Janey Dudley, Humza Siddiqi, and Anna Wani

Background

Sleep plays a foundational role in an infant's development and overall well-being, influencing critical processes such as brain maturation, physical growth, and emotional regulation [1, 2]. Unlike adults, infants experience distinct and rapidly changing sleep patterns, including shorter sleep cycles, frequent night awakenings, and a greater proportion of rapid eye movement (REM) sleep, all of which reflect the unique developmental needs of early life. For caregivers, understanding these evolving patterns is essential in creating a supportive sleep environment and fostering healthy sleep habits from the start [3–5]. This chapter explores the science of infant sleep, common challenges such as night waking and inconsistent sleep schedules, and offers evidence-based strategies to promote better sleep quality. Drawing from the latest research and clinical guidelines, it aims to empower caregivers and clinicians alike to support more restful nights—for both infants and their families.

Newborn Sleep Patterns

Newborns have a distinct sleep architecture that variers significantly from adults. They typically sleep around 16–18 hours a day. However, this sleep is fragmented into short periods of 2–4 hours throughout the day and night [6]. This pattern is mainly because of their small stomach capacity, necessitating frequent feedings [7].

Recent studies suggest that these brief sleep cycles help newborns adjust to their new extra-uterine environment [8]. Newborns spend a significant portion of their sleep in REM sleep, accounting for about 50% of their total sleep time [6, 9]. REM sleep is lighter and more active and is described to be pivotal for brain development and the processing of sensory input [10]. During this phase, newborns may exhibit twitches, smiles, or other small movements, indicating active brain processing.

The Development of Sleep Cycles

As infants grow, their sleep cycles begin to lengthen and consolidate. By around 3 to 4 months, most of the infants start to establish a more predictable sleep-wake pattern. At this age, their total sleep time gradually decreases to about 14–15 hours per day [11]. During this phase, infants spend more time in the deeper stages of non-REM sleep, during which a maximum of the restorative changes occur. As this transitions into longer periods of non-REM sleep, physical growth and immune function increases [12]. Around 6 months, many infants can sleep for longer stretches at night, sometimes up to 6–8 hours, providing some much-needed relief to sleep-deprived parents [4]. However, it's important to note that there exist exceptions to this norm of timeline, and there is considerable individual variation in sleep patterns [13].

The Importance of Sleep for Development

Sleep affects infant's growth and development in multiple ways. The growth hormone produced during sleep aids in physical development [14]. Adequate sleep supports brain development by facilitating the formation of neural connections that are crucial for learning and memory consolidation [15–17]. Recent research found associations between infant sleep quality and cognitive and behavioral outcomes, emphasizing the role of early sleep in long-term development [18]. In addition to physical and cognitive benefits, sleep also impacts emotional regulation. Infants with well-rested sleep tend to be more alert, content, and better able to interact with their environment. Conversely, infants with sleep deprivation or poor sleep quality may be fussier, harder to soothe, and more prone to emotional distress [19]. Newer studies suggest that poor sleep in infancy may be linked to an increased risk of emotional and behavioral issues in childhood [20].

Common Sleep Challenges

A. Night Wakings

Frequent night awakenings are one of the most common sleep challenges parents face. It's normal for infants to wake up several times during the night, particularly in the first few months [7]. These awakenings may be due to hunger, discomfort, or simply the need for comfort and soothing. Recent evidence indicates that night wakings are a normal part of sleep development and reflect the infant's need to synchronize with their caregiver [21, 22]. As infants develop, their ability to self-soothe and fall back asleep without parental intervention becomes an important skill. While some infants naturally develop this ability around 4–6 months, others may need more consistent support and motivation from their caregivers [23]. Sleep training methods that promote self-soothing have been shown to reduce night wakings and improve overall sleep quality [24].

B. Sleep Regressions

Sleep regressions are times when a baby's sleep patterns abruptly change, leading to more frequent night wakings and shorter naps. These regressions usually happen around 4 months, 8–10 months, and 12 months and are often linked to developmental milestones like rolling over, crawling, or standing [5, 16]. Research suggests that these regressions may be connected to cognitive and motor skill development, which can temporarily interfere with sleep [25, 26].

Although it may be tempting to introduce new sleep aids or routines during these sleep regression phases, maintaining established sleep practices is generally advisable. Regressions are typically temporary, lasting a few weeks before sleep returns to a more predictable pattern [27]. A consistent bedtime routine and sleep environment can help minimize the impact of these regressions.

C. Separation Anxiety

Around 6–9 months, infants develop a sense of object permanence, understanding that people and objects continue to exist even when out of sight [5, 26]. This cognitive leap can lead to separation anxiety, making it more challenging for infants to settle down for sleep without their caregiver nearby. Recent studies have shown that separation anxiety during this stage is a normal developmental process and often coincides with significant social and emotional growth [28]. To ease separation anxiety, establishing a consistent bedtime routine and providing reassurance through comforting words or a favorite transitional object, like a soft blanket or stuffed animal, can be helpful [22]. Research indicates that gradual sleep training methods, which involve gradually increasing the time between parental interventions, can be effective in reducing separation anxiety [29, 30].

Establishing Healthy Sleep Habits

1. Creating a Sleep-Friendly Environment

Creating a conducive sleep environment is crucial for helping an infant sleep well. The sleep space should be quiet, dark, and cool [1, 31]. Blackout curtains can block light, and a white noise machine can mask household sounds, enhancing the sleep environment. Additionally, the crib should be free from loose bedding, pillows, or toys to minimize the risk of suffocation and sudden infant death syndrome (SIDS) [31]. Recent guidelines also stress placing infants on their backs to further lower SIDS risk [32]. Establishing a consistent sleeping area, whether it's a crib in the parents' room or a separate nursery, helps signal to the baby that it's time to sleep. Consistency in the sleep environment fosters a sense of security and routine, which can be especially beneficial during transitions or developmental changes [33].

2. Developing a Bedtime Routine

A predictable bedtime routine is used to signal to the baby that it's time to wind down and prepare for sleep. Activities such as a warm bath, gentle rocking, reading a short book, or singing a lullaby might be included in this routine [16, 24]. The routine should be made calm and soothing, with stimulating activities avoided to prevent making it harder for the baby to settle down. Recent studies have confirmed that improved sleep outcomes in infants and young children are associated with a consistent bedtime routine [17]. By maintaining a consistent bedtime routine and keeping it around the same time each night, the transition from wakefulness to sleep is facilitated for the baby. Over time, this routine becomes a cue that helps the baby recognize that it's time to sleep, leading to better sleep quality and duration.

3. Encouraging Self-Soothing

Teaching an infant to self-soothe is an important step in helping them sleep through the night. This doesn't mean leaving them to cry indefinitely but rather giving them the opportunity to learn how to fall asleep on their own [21, 23]. Start by putting the baby down to sleep while they are drowsy but still awake so they can associate falling asleep with being in their crib. If the baby wakes up during the night, give them a moment to see if they can settle back to sleep on their own. If they continue to cry, use a gentle and consistent approach to offer comfort, such as patting or shushing, without immediately picking them up. Over time, they will learn that they can soothe themselves back to sleep [16]. Recent evidence suggests that sleep training methods that encourage self-soothing are effective in reducing sleep problems and improving parental well-being [29].

When to Seek Help

While many variations in infant sleep and breathing are normal, it is important to recognize when further evaluation is warranted. Irregular breathing patterns and rapid-cycle breathing are often normal in young infants, especially during REM sleep. However, signs such as color changes (e.g., turning blue or pale), limpness, or concerning events such as Apparent Life-Threatening Events (ALTE) or Brief Resolved Unexplained Events (BRUE) should prompt immediate medical attention and further investigation. Though rare, conditions like Congenital Central Hypoventilation Syndrome (CCHS) must be considered in infants with persistent abnormal breathing during sleep. Additionally, gastroesophageal reflux (GER) is common in infancy and can lead to sleep disruption, particularly when associated with discomfort or feeding difficulties. Providers should help families distinguish between typical infant behaviors and signs that require clinical assessment, ensuring that sleep disturbances are not overlooked when they may signal a more serious underlying condition. Although many sleep challenges are a normal part of infant development, there are situations where seeking help from a healthcare professional is necessary. If the baby consistently struggles with sleep, appears excessively irritable, or there are concerns about their growth and development, consulting a pediatrician is crucial [31]. Recent guidelines highlight the importance of addressing sleep problems early to prevent potential long-term consequences [32].

Sleep issues can sometimes indicate underlying medical conditions such as reflux, allergies, or sleep apnea. Early intervention can address these issues and support better sleep for both the baby and the parents. New research also suggests that certain behavioral and the parents [33]. New research also suggests that certain behavioral and environmental interventions can significantly improve sleep quality for infants [21].

Clinical Pearls on Infant Sleep

1. Newborn Sleep: Newborns typically sleep 16–18 hours a day in short cycles, with a significant portion of their sleep in REM, which is crucial for brain development.
2. Sleep Development: By 3–4 months, infants begin to consolidate sleep, with longer non-REM sleep periods that promote physical growth and immune function.
3. Sleep Benefits: Adequate sleep supports physical growth, brain development, and emotional regulation, helping infants stay alert, content, and better interact with their environment.
4. Common Challenges:
    (a) Night Wakings: Frequent awakenings are normal in the early months, but infants can learn to self-soothe by 4–6 months.
    (b) Sleep Regressions: Sleep disruptions linked to developmental milestones are common, but maintaining consistent sleep routines can help manage them.
    (c) Separation Anxiety: Around 6–9 months, infants may experience separation anxiety, and gradual sleep training can assist in easing this.
5. Healthy Sleep Habits:
    (a) Environment: A quiet, dark, and safe sleep space is essential for promoting better sleep.
    (b) Routine: Establishing a consistent and soothing bedtime routine helps infants transition from wakefulness to sleep.
    (c) Self-Soothing: Encourage infants to fall asleep drowsy but awake to develop self-soothing skills.
6. When to Seek Help: If sleep issues persist or if there are concerns about the infant's growth or development, it's important to consult. Sleep Medicine for further evaluations including a sleep study, and sometimes further clinical evaluation for GERD and ENT are warranted. Acuity of symptoms should also be addressed.

Conclusion

Infant sleep is essential for growth, brain development, and emotional regulation. Understanding sleep patterns, such as REM sleep and night wakings, helps caregivers support healthy sleep habits. Addressing challenges like sleep regressions and separation anxiety, while fostering a consistent sleep environment and routine, can improve sleep quality. While sleep issues are common, seeking professional help is important if problems persist or if there are concerns about development. Healthy sleep in infancy is crucial for overall well-being and future development.

References

1. Mindell JA, Owens JA. A clinical guide to pediatric sleep: diagnosis and management of sleep problems. Lippincott Williams & Wilkins; 2015.
2. Grigg-Damberger MM. The visual scoring of sleep in infants 0–2 months of age. J Clin Sleep Med. 2009;5(2):125–38.
3. Hale L, Guan S. Screen time and sleep among school-aged children and adolescents: a systematic literature review. Sleep Med Rev. 2015;21:50–8.
4. American Academy of Pediatrics. Sleep and your baby: 6 to 12 months. https://healthychildren.org. Accessed 24 Nov 2024.
5. Children's Hospital of Philadelphia. Healthy sleep habits. https://www.chop.edu. Accessed 24 Nov 2024.
6. Galland BC, Mitchell EA, Currie J. Sleep and infancy. J Sleep Res. 2012;21(2):247–57.
7. Paul IM, Gauthier JA, Paquin R. Sleep patterns and developmental milestones. Pediatrics. 2021;147(2):e20200317.
8. Grigg-Damberger MM. Normal sleep in infants and children. In: Sheldon SH, Ferber R, Kryger MH, editors. Principles and practice of pediatric sleep medicine. Elsevier; 2016. p. 35–46.
9. Vollmer MA, Thomas M, Meyer JS. The role of REM sleep in infant development. Sleep Med Rev. 2021;53:35–45.
10. Morgenthaler TI, Owens J, Alessi C, et al. Practice parameters for behavioral treatment of bedtime problems and night wakings in infants and young children. Sleep. 2006;29(10):1277–81.
11. Harvard Medical School, Division of Sleep Medicine. Healthy sleep in children: How many hours does your child need? https://healthysleep.med.harvard.edu. Accessed 24 Nov 2024.
12. Hysing M, Pallesen S, Stormark KM. Sleep patterns in infants. Sleep Med Rev. 2022;26(1):19–32.
13. Bruni O, Novelli L, Ferri R. Growth hormone release and sleep. Sleep Med Rev. 2008;12(1):31–45.
14. Walker MP. The role of sleep in cognition and emotion. Ann N Y Acad Sci. 2009;1156:168–97.
15. Mindell JA, Kuhn B, Lewin DS, Meltzer LJ, Sadeh A. Behavioral treatment of bedtime problems and night wakings in infants and young children. Sleep. 2006;29(10):1263–76.
16. Baby Sleep Made Simple. 12-month sleep regression survival guide. https://www.babysleepmadesimple.com. Accessed 24 Nov 2024.
17. Tobon A, Hill CM, Santoro M, De Gennaro L, Miccoli M. Infant sleep quality and its relationship with cognitive and behavioral outcomes later in life. Child Dev. 2023;94(2):345–60.
18. Bates JE, Viken RB, Alexander DB, Kindlon D. Sleep and behavior in early childhood. J Child Psychol Psychiatry. 2002;43(2):135–44.
19. Tham SK, Ponsonby AL. Infant sleep and later behavioral problems. J Child Psychol Psychiatry. 2021;62(4):448–57.
20. Crivillaro RP, Santos LS, Barbosa TS. Behavioral and environmental interventions for infant sleep. Pediatr Sleep Med. 2023;19(3):205–16.

21. Sadeh A. Cognitive-behavioral treatment for childhood sleep disorders. Clin Child Psychol Psychiatry. 2005;10(3):303–15.
22. Ferber R. Solve your child's sleep problems. Simon & Schuster; 1985.
23. Gradisar M, Blunden S, Flay M. Sleep training for infants and young children. Sleep Med Clin. 2016;11(3):419–30.
24. St James-Roberts I. Sleep regressions in infancy. Sleep Med Rev. 2018;37:35–46.
25. Piaget J. The construction of reality in the child. Basic Books; 1954.
26. MadeForMums. Object permanence: What age and how do babies learn it? Updated October 26, 2023. https://www.madeformums.com. Accessed 24 Nov 2024.
27. Pluess M, Belsky J. The development of object permanence and separation anxiety. Child Dev Perspect. 2018;12(2):123–8.
28. Hiscock H, Wake M. Effectiveness of sleep training methods. Pediatrics. 2020;145(4):e20193609.
29. Lange A, Weber A, Fuchs L, et al. Effectiveness of gradual sleep training in infants: a meta-analysis. Sleep Med Rev. 2023;64:101635.
30. Moon RY. Task Force on Sudden Infant Death Syndrome. SIDS and other sleep-related infant deaths: evidence base for 2016 updated recommendations for a safe infant sleeping environment. Pediatrics. 2016;138(5):e20162938.
31. American Academy of Pediatrics. Policy statement: SIDS and other sleep-related infant deaths: updated 2022 recommendations for a safe infant sleeping environment. Pediatrics. 2022;150(6)
32. Adair J, Mortimer JT. Infant sleep environments and developmental outcomes: evidence on the importance of a consistent sleeping area. Child Dev Perspect. 2023;17(3):145–52.
33. Owens JA, Spirito A, McGuinn M, Nobile C. Sleep habits and sleep disturbance in elementary school-aged children. J Dev Behav Pediatr. 2003;21(1):27–36.

Toddler Sleep

6

Liz Lezama, Shan Luong, Victoria Udezi, Esha Banwatt, and Elisa Basora-Rovira

Introduction

Sleep is an active process that continuously evolves throughout life. Age is the most well-established determinant of sleep duration, with total sleep time decreasing from infancy through adulthood. Cognitively, toddlerhood represents a phase of rapid changes and challenges for parents and providers. During these years, children displayed the most dramatic changes in neurocognitive development, and sleep is not the exception. Toddlers should sleep 11–14 hours in 24 hours, including naps. During this phase, they experienced consolidation of daytime and nighttime sleep, displaying continuous sleep of about 10 hours with fewer interruptions and 1–2 well consolidated daytime naps.

Even though toddlers have a similar sleep pattern compared to adults, cognitively things are different. Understanding normal neurodevelopment during this age is key to understanding the changes and challenges that toddlers experience.

L. Lezama · E. Basora-Rovira (✉)
Department of Pediatric Pulmonology and Sleep Medicine, UT Southwestern, Dallas, TX, USA
e-mail: liz.lezama@utsouthwestern.edu; elisa.basorarovira@utsouthwestern.edu

S. Luong
Department of Internal Medicine – Pulmonary Division, UT Southwestern, Dallas, TX, USA
e-mail: shan.luong@utsouthwestern.edu

V. Udezi · E. Banwatt
Department of Community and Family Medicine, UT Southwestern, Dallas, TX, USA
e-mail: victoria.udezi@utsouthwestern.edu; esha.banwatt@utsouthwestern.edu

© The Author(s), under exclusive license to Springer Nature Switzerland AG 2025
A. Wani, I. S. Khawaja (eds.), *Sleep Disorders in Children*,
https://doi.org/10.1007/978-3-031-92166-7_6

Normal Sleep in Toddlers and Night-Time Awakenings

According to the American Academy of Sleep Medicine (AASM), patients between ages of 1 and 3 years of age should have 11–14 hours of sleep including naps. Sleep during early childhood is often influenced by a constellation of biological, behavioral, and cultural factors, amongst others. Studies have found an association between sleep quality as reflected by the frequency of night-time awakenings and cognitive abilities in young children, in particular, poorer cognitive performance in toddlers with more frequent night-time awakenings. The prevalence of night-time awakenings among children 0–3 years of age ranges from 20% to 60% depending on the age and defining criteria used; this could lead to fragmented sleep and compromise the sleep quality in young children. However, few studies have investigated the impact of night-time awakenings in relation to cognitive development in toddlers.

Insomnia

Insomnia is the difficulty to initiate sleep, consolidation or quality of sleep, resulting in daytime impairment. According to the International Classification of Sleep Disorders (ICSD-3), chronic insomnia is diagnosed when the symptoms occur at least three times per week, and the sleep disturbances with the associated daytime symptoms are present for at least 3 months. For symptoms less than 3 months, the diagnosis of acute short-term insomnia is used. The prevalence of insomnia in early childhood has been reported in a wide range from 10% to 50%. Studies have found that insomnia is the most reported sleep issue in pediatric sleep clinics with 41.4% reporting at least one symptom of insomnia.

Pediatric insomnia will manifest with neurobehavioral symptoms such as hyperactivity, problems with concentration, academic problems, or fatigue. Nonetheless night-time awakenings are reported as one of the most common symptoms, especially in infants and toddlers; bedtime resistance is also seen in about 10–15% of toddlers. Unfortunately, insomnia and other sleep disorders in children are underdiagnosed.

Types of Pediatric Insomnia

1. *Sleep Onset Association*: children who refuse to sleep because they need a specific behavioral pattern or external factors (e.g., objects, specific person, rocking) to fall asleep; it is characterized by multiple night-time awakenings. It is very common in young infants and toddlers.
2. *Limit-Setting*: Children who have difficulty following bedtime rules (tantrums) or tend to oppose their parents, especially at bedtime; this type is often observed in pre-school and school-aged children.
3. *Combined*: Children who displayed symptoms of the two subtypes.

Etiology

Intrinsic and extrinsic factors are associated with increased risk of pediatric insomnia, as well as neuropsychological developmental stage of the child. There can be age-specific causes for insomnia, however, it should be noted that overlaps between age groups can occur (Table 6.1). Many medical conditions have been known to be disruptive with sleep continuity and architecture resulting in misalignment of the circadian clock favoring the development of insomnia. Additionally, neurobehavioral conditions may foster dysfunctional sleep/wake cycle. On the other hand, extrinsic factors are mainly dictated by environmental aspects or caregiver responses that promote the development and perpetuation of insomnia.

Assessment and Diagnosis

Knowing the developmental norms is important when considering insomnia as a diagnosis. In toddlers, night wakings require parental involvement, are more likely related to sleep onset association insomnia, and are associated with less consolidated sleep. It is also important to remember that pediatric insomnia presents differently than in adolescents and adults, mainly characterized by behavioral changes rather than daytime sleepiness; therefore, it is essential to have a structured approach for an accurate assessment of this disorder. Diagnostic tools must include a comprehensive clinical interview with focus on sleep history, developmental history, sleep diaries, the evaluation of comorbidities, and the collection of objective sleep data.

The clinician should evaluate bedtime routines, sleep/awake patterns, sleep environment, difficulties with sleep initiation or maintenance, abnormal movements or behaviors during sleep, and daytime impairment. The interview should also include

Table 6.1 Etiology/risk factors of insomnia in infants and toddlers

Age Range	Causes/Risk factors
Infants	Sleep onset association disorder (SOAD)
	Infant colic
	Gastroesophageal reflux
	Milk allergy
	Infections
	Medical or psychiatric disorder in a caregiver (e.g., Maternal anxiety)
Toddlers	Sleep-onset association disorder (SOAD)
	Limit-setting sleep disorder
	Parental separation anxiety
	Fear
	Prolonged naps
	Infections
	Neurobehavioral conditions (Autism, ADHD)
	Medical or psychiatric disorder in a caregiver (e.g., Maternal anxiety)
	Inconsistent parenting styles
	Distracting bedroom environment
	Excessive screen time

a description of previous interventions, including medications or behavioral strategies. As mentioned above, sleep diaries are a valuable strategy for the diagnosis and monitoring of pediatric insomnia; these can be completed 2 weeks prior or after the initial evaluation and should include parental report on sleep patterns and characteristics. This data will guide the targets for intervention and tracking of the progress.

Using technology to gather sleep data is common to assess pediatric insomnia. Actigraphy is a frequently used tool to evaluate insomnia in children; it involves wearing a wristwatch-like device to monitor movement throughout the day and night. This device is typically worn for 2 weeks, after which data can be processed and analyzed generating a summary of sleep and wake patterns. Some literature has compared actigraphy with sleep diaries, learning that actigraphy is more accurate in estimating sleep onset latency and total sleep time when compared to parental reports. Typically, actigraphy is often preferred over more invasive tools (e.g., Polysomnography) which are usually not indicated for the assessment of pediatric insomnia.

Social Determinants of Health and Sleep in Toddlers

Mechanisms of sleep deficiency in racial/ethnic minorities are not well understood but include environmental factors such as food and house insecurity, lack of education, and unsafe neighborhoods. Children from low-income Black families experienced less night-time sleep time, later bedtimes, and poorer sleep routines compared to children from Caucasian families with high socioeconomic status. Understanding the maternal, social, and environmental factors that place toddlers in minority, low-income families at increased risk for inadequate sleep and poorer health outcomes might facilitate the development of strategies to reduce sleep disparities and improve outcomes.

Differential Diagnosis

Sleep-Disordered Breathing

Sleep-disordered breathing (SDB) is defined as the disruption of normal respiratory patterns and ventilation during sleep [1]. This causes a child to stop breathing briefly or to have shallow breathing while sleeping. It has been estimated by the American Academy of Pediatrics that 1.2–5.7% of children are affected by SDB caused by obstructive sleep apnea (OSA) alone [2]. There are also concerns that as prevalence of childhood obesity increases, these figures are likely to be higher. SDB can contribute to behavioral and physical health concerns in children. The effects of SDB on the health of children include consequences of OSA on growth and development, activity levels, learning, and behavior. Fatigue, irritability, and mood changes may also be present. Insufficient sleep can also lead to the development or exacerbation of chronic diseases (Table 6.2) [2].

Table 6.2 American Academy of Pediatrics 2012 Guideline Recommendations for the Diagnosis and Treatment of OSA in Children and Adolescents [2]

1. All children/adolescents should be screened for snoring.
2. Polysomnography should be performed in children/adolescents with snoring and symptoms/signs of OSA; if polysomnography is not available, then alternative diagnostic tests or referral to a specialist for more extensive evaluation may be considered.
3. Adenotonsillectomy is recommended as the first-line treatment of patients with adenotonsillar hypertrophy.
4. High-risk patients should be monitored as inpatients postoperatively.
5. Patients should be reevaluated postoperatively to determine whether further treatment is required. Objective testing should be performed in patients who are high risk or have persistent symptoms/signs of OSA after therapy.
6. Continuous positive airway pressure is recommended as treatment if adenotonsillectomy is not performed or if OSA persists postoperatively.
7. Weight loss is recommended in addition to other therapies in patients who are overweight or obese.
8. Intranasal corticosteroids are an option for children with mild OSA in whom adenotonsillectomy is contraindicated or for mild postoperative OSA.

OSA obstructive sleep apnea

Children and their caregivers may also bear physical and mental health effects as disrupted sleep can be a major stressor on the household. SDB may be diagnosed through polysomnography (PSG) after an adequate history and physical examination. Adenotonsillectomy is the primary treatment for most children with obstructive sleep apnea due to adenotonsilar hypertrophy. Pediatric clinicians are advised to screen for, provide education, and address, SDB in their patients.

Medical Insomnia

Sleep disorders in toddlers may be related to acute and chronic medical conditions. Healthcare practitioners should be familiar with medical diagnosis that may be associated with sleep disturbance in toddlers. Toddlers presenting with diverse medical conditions may experience symptoms such as pain or discomfort that can affect sleep. These include but are not limited to:

- *Allergies:* About 40% of children experience allergic rhinitis in the United States [3]. Toddlers with seasonal or chronic allergies develop symptoms such as congestion, pruritus, coughing, and sneezing that can affect sleep leading to insomnia. Children older than 3 years old can be treated with intranasal steroid sprays. However, some antihistamine treatments, especially combined with decongestants, may worsen sleep.
- *Upper Airway Infections:* Acute and chronic infections with their respective symptoms (e.g., fever, pain, irritability, and appetite changes) may cause prolonged or repeated wakings leading to daytime sleepiness and disrupted nocturnal sleep. Sleep disruption caused by acute infection should resolve shortly after adequate treatment while chronic infections (e.g., chronic otitis media) may lead to chronic discomfort and problematic sleep patterns.

- *Respiratory conditions (Asthma, Cystic Fibrosis):* Sleep disturbances can be noted due to symptoms such as nighttime cough or disturbed breathing. Also, medications used to treat these conditions may also result in insomnia.
- *Atopic Dermatitis:* Toddlers with atopic dermatitis may experience sleep disruption related to symptoms such as itching or medications used to treat the condition.
- *Teething:* More than 80% of infants and toddlers experience nocturnal sleep disturbances [4]. Pain and discomfort associated with teething can make it difficult for infants and toddlers to sleep during the night. Symptoms may peak prior to tooth eruption and may continue for a few days after. Symptomatic care may provide relief and improve sleep patterns.

Periodic Limb Movement Disorder (PLMD)

PLMD is characterized by repetitive stereotypic movements in the legs during sleep, more rarely in the arms [5]. It is a diagnosis made from polysomnography that shows >= 5 periodic limb movements per hour of sleep and is associated with clinically significant sleep disturbance or negative impact on daytime functioning [6]. This sleep disorder diagnosis is made in the absence of other contributing sleep, medical or psychiatric disorder or medications. Sleep disturbance is usually defined as either sleep onset problems where sleep latency is >20 minutes that is occurring twice or more per week or sleep maintenance problems of >= 2 full arousals per night that is occurring twice or more per week [7]. Impaired daytime functioning is usually gauged through parent-reported pediatric sleep questionnaires including daytime fatigue, hypersomnia, hyperactivity, irritability, inattention, aggressive behavior, and anxiety [5]. The hypothesized pathophysiology of the disorder is thought to be related to the impairment of dopaminergic transmission also seen in restless legs syndrome (RLS). The prevalence of PLMD at a referred sleep center was reported at 1.2% in children without other comorbidities [5]. The prevalence of PLMS tends to increase with age, to around 3.9% in adults [7]. Treatment usually involves oral or intravenous iron repletion to target serum ferritin level > 50 ng/mL.

Restless Legs Syndrome (RLS)

RLS is a type of movement disorder characterized by an uncomfortable feeling in the legs and/or arms and can be disruptive to sleep. RLS diagnosis is made clinically with the tetrad of "urge to move the legs, worsening of symptoms during rest, worsening of symptoms in the evening and improvement or resolution of symptoms after movement" [5]. Children should be asked if there is an "urge to move" using age-appropriate prompts. The prevalence of RLS ranges from 2% to 4% of normally developing children and adolescents through a questionnaire completed by mother [6]. There can be a genetic component to RLS. Common symptoms include excessive daytime sleepiness, disrupted sleep, mood changes, attention deficit, and other behavioral problems such as irritability, aggression, hyperactivity [5]. RLS and PLMD are similar but are considered mutually exclusive. Treatment includes non-pharmacologic approaches such as sleep hygiene, diet control, physical activity in the form of stretching exercises before bedtime [5]. There are no Food and Drug Administration (FDA) approved medications for the treatment of RLS in children. Pharmacologic interventions are usually off-label use of iron supplement to target serum ferritin level >= 50 ng/mL. Alternatives or add-on medications for those not responsive to iron supplementation include dopamine agonists such as pramipexole, gabapentin, clonazepam, and clonidine [5].

Restless Sleep Disorder (RSD)

The diagnosis of RSD applies to children aged 6–18 [8]. Diagnostic criteria for RSD were established recently in 2019, therefore, we do not yet have robust data regarding its prevalence in the general population. Approximately 7.7% of pediatric patients referred to a sleep clinic were found to have RSD [9]. There is a higher prevalence of RSD among children with ADHD and parasomnias [5]. The following are essential diagnostic criteria for RSD: a complaint of restless sleep from patient, bedmate, or caregiver, Large whole-body movements which occur during sleep $\geq 3\times$ weekly for ≥ 3 months, and clinically significant impairment in daytime functioning and behavior. Video polysomnography is required for diagnosis and must reveal a total movement index of ≥ 5 movements per hour. Finally, a patient's symptoms should not be better explained by an alternative disorder [8]. Currently, the proposed pathophysiology of RSD is centered on iron deficiency altering dopaminergic motor pathways and contributing to restlessness [5]. In children suspected to have RSD, obtaining a vPSG study, iron studies including ferritin, and reinforcing sleep hygiene are appropriate initial steps. Children with a ferritin level ≤ 50 mcg/L will likely benefit from iron supplementation (15 mg/ kg IV once or 325 mg ferrous sulfate PO daily for 8 weeks) [10].

Workup

When a sleep disorder is identified in a toddler, taking a thorough history and physical exam will guide toward the etiology and differential diagnosis. At this age, the most common sleep disorders are of behavioral origin. Although, other sleep disorders may be present, as described above. Parents and/or caregivers can help the provider with a sleep log which can be used to track and understand sleep patterns during sleep. Actigraphy is another test that is available to objectively measure the quantity of sleep [11, 12]. It is worn on the wrist for a period of 2–3 weeks and provides estimates on the total time spent in bed, sleep latency, total sleep duration, and sleep efficiency.

In cases of restless legs syndrome (RLS) or restless sleep disorder, iron deficiency has been a contributor to the symptoms. In most of the cases of RLS, ferritin levels less than 50 ng/ml have been found, with improvement on RLS symptoms after iron supplementation in children [13]. Other medical conditions such as renal failure, hypothyroidism, Diabetes Mellitus, polyneuropathy, vitamin B12 deficiency, and magnesium deficiency have been associated with RLS symptoms.

If sleep disordered breathing is suspected, the gold standard test is polysomnography. This is an overnight test that records different sleep stages, breathing with possible breathing pauses, movements during sleep, among other values.

Treatment

Behavioral interventions are the most important treatment for behavioral insomnia of childhood. The most efficacious therapies include bedtime routines, systematic extinction and its variants, bedtime fading, and positive reinforcement. Sleep-related interventions have been shown to improve other variables in children, such as daytime behavior, and the well-being of parents [14]. It is important to emphasize that the child's safety with an adequate sleep environment is the mainstay in achieving therapy success. Socioeconomic factors that affect sleep and treatment need to be taken into consideration when establishing behavioral interventions with parents and/or caregivers.

Bedtime routines: Educating parents and/or caregivers on how to establish a consistent bedtime routine is key in all types of behavioral insomnia. It includes a set of activities every night 30–60 minutes before bedtime.

- Extinction variants: These therapies help to extinguish the child's need to have the presence of a parent in order to fall asleep. Unmodified extinction ("crying it out") is a therapy that involves placing the toddler to bed drowsy and then leaving the room without going back in. The child learns to fall asleep on their own and self-soothe. This is a very successful approach, although some families don't tolerate it. Graduate extinction is another therapy that involves removing the parental presence in a more gradual way within 1 to 2 weeks. The child will be placed in bed drowsy, and the parent will leave the room with progressively longer waiting periods before checking on the child. Other gradual techniques are available and appropriate for families depending on the level of tolerance and acceptability [15].
- Bedtime fading: This therapy involves moving the set bedtime to the child's circadian preference and gradually advancing it over a period of several weeks.
- Positive reinforcement: Toddlers, preschoolers, and some older children may respond well to reinforcement strategies to target behaviors such as staying in bed all night and reducing the number of calls to parents. Sticker charts or token economy are examples in which the child will earn a sticker or token and later achieve a goal or exchange something of a child's interest.

Many other non-pharmacological strategies may improve sleep at this age. Some of them includes avoid napping 4 hrs. before desired bedtime; no screentime 30–60 minutes before bedtime; reduce emotional and mental stimulation at bedtime; sleep in a bedroom when possible on a mattress with a pillow; having a cool (68°–72 °F) and quiet sleep environment; consider blackout curtains and sound machines to reduce disturbances, avoid caffeinated drinks.

Pharmacologic Management of Insomnia

Pharmacological management is not routinely recommended in toddlers and will not substitute healthy sleep hygiene and non-pharmacological strategies. There is no drug approved by the US Food and Drug Administration (FDA) for the treatment of insomnia in children. There are certain populations (e.g., children with neurologic conditions, psychiatric conditions, autism) that some sleep aids are recommended as and adjuvant to therapies due to clinical experience or small studies.

Melatonin is a hormone produced by the pineal gland in response to darkness that regulates the circadian rhythm. Its synthetic form is available in the United States as "over the counter" with minimal Food and Drug Administration (FDA) regulations. Limited data is available on the use of melatonin in children supporting its use for sleep onset insomnia. However, the dosing of melatonin is based on sleep disorder and child's age. Some studies showed that melatonin is overall well-tolerated with minimal side effects. However, long-term studies are not available. Due to limited data available for this population, it is recommended to discuss with patient's primary or sleep provider in regards to the use of melatonin in toddlers.

Clinical Consequences of Insufficient Sleep in Toddlers

Overall, studies have shown that insufficient sleep in children may be related to daytime sleepiness, irritability, emotional and behavioral problems, learning difficulties, poor academic performance, sports injury, and motor vehicle crashes in teenagers. However, long-term studies in toddlers have not been performed. Having sufficient sleep is key in the pediatric population due to ongoing body and brain development to support physical growth, cognitive development, emotional regulation, and immune function.

Toddler Sleep Case
A 2-year-old female with no significant past medical history comes to clinic accompanied by mother for restless sleep evaluation. Mother reports that she is very active before going to bed and will ask for her electronic tablet while on her bed. She will start crying and fighting if she does not get her electronic tablet but will eventually

fall asleep. She usually goes to bed at 10:00 pm but falls asleep around 12–2:00 am and wakes up at 12:00 pm. She takes a nap for one hour, usually at 1:00 pm. Her sleep routine is variable. She sleeps with mother in her bed. She tosses and turns throughout the night. Upon further diagnostic work-up ferritin levels were obtained due to concerns for restless sleep. Ferritin levels resulted in 25 ng/ml. Iron therapy initiated with improvement in restlessness; however, the patient continued to display the need for technological devices and maternal presence to fall asleep.

References

1. Gipson K, Lu M, Kinane TB. Sleep-disordered breathing in children. Pediatr Rev. 2019a;40(1):3–13. https://doi.org/10.1542/pir.2018-0142. Erratum in: Pediatr Rev. 2019 May;40(5):261. https://doi.org/10.1542/pir.405261. PMID: 30600274; PMCID: PMC6557418.
2. Gipson K, Mengdi L, Bernard Kinane T. Sleep-disordered breathing in children. Pediatr Rev. 2019b;40(1):3–13. https://doi.org/10.1542/pir.2018-0142.
3. Meltzer EO, Blaiss MS, Derebery MJ, Mahr TA, Gordon BR, Sheth KK, Simmons AL, Wingertzahn MA, Boyle JM. Burden of allergic rhinitis: results from the pediatric allergies in America survey. J Allergy Clin Immunol. 2009;124(3 Suppl):S43–70. https://doi.org/10.1016/j.jaci.2009.05.013. Epub 2009 Jul 9. PMID: 19592081.
4. Memarpour M, Soltanimehr E, Eskandarian T. Signs and symptoms associated with primary tooth eruption: a clinical trial of nonpharmacological remedies. 2015. https://pubmed.ncbi.nlm.nih.gov/26215351/.
5. DelRosso LM, Mogavero MP, Bruni O, Ferri R. Restless legs syndrome and restless sleep disorder in children. Sleep Med Clin. 2023;18(2):201–12. https://doi.org/10.1016/j.jsmc.2023.01.008.
6. Hornyak M, Feige B, Riemann D, Voderholzer U. Periodic leg movements in sleep and periodic limb movement disorder: prevalence, clinical significance and treatment. Sleep Med Rev. 2006;10(3):169–77. https://doi.org/10.1016/j.smrv.2005.12.003.
7. DelRosso LM, Picchietti DL, Sharon D, et al. Periodic limb movement disorder in children: a systematic review. Sleep Med Rev. 2024;76:101935. https://doi.org/10.1016/j.smrv.2024.101935.
8. DelRosso LM, Ferri R, Allen RP, et al. Consensus diagnostic criteria for a newly defined pediatric sleep disorder: restless sleep disorder (Rsd). Sleep Med. 2020;75:335–40. https://doi.org/10.1016/j.sleep.2020.08.011.
9. DelRosso LM, Ferri R. The prevalence of restless sleep disorder among a clinical sample of children and adolescents referred to a sleep centre. J Sleep Res. 2019;28(6):e12870. https://doi.org/10.1111/jsr.12870.
10. DelRosso LM, Picchietti DL, Ferri R. Comparison between oral ferrous sulfate and intravenous ferric carboxymaltose in children with restless sleep disorder. Sleep. 2021;44(2):zsaa155. https://doi.org/10.1093/sleep/zsaa155.
11. Smith MT, et al. Use of actigraphy for the evaluation of sleep disorders and circadian rhythm sleep-wake disorders: an American Academy of Sleep Medicine systematic review, meta-analysis, and GRADE assessment. J Clin Sleep Med. 2018;14(7):1209–30. https://doi.org/10.5664/jcsm.722842. PMID 29991438.
12. Morgenthaler T, et al. Practice parameters for the use of actigraphy in the assessment of sleep and sleep disorders: an update for 2007. Sleep. 2007;30(4):519–29. https://doi.org/10.1093/sleep/30.4.519. PMID 17520797.

13. DelRosso LM, Mogavero MP, Bruni O, Ferri R. Restless legs syndrome and restless sleep disorder in children. Sleep Med Clin 2023c;18(2):201–212. doi: https //doi.org/10.1016/j.jsmc.2023.01.008. PMID: 37120162.
14. Wolfson A, Lacks P, Futterman A. Effects of parent training on infant sleeping patterns, parents' stress, and perceived parental competence. J Consult Clin Psychol 1992;60(1):41–48. doi: https://doi.org/10.1037//0022-006x.60.1.41. PMID: 1556284.
15. Gradisar M, Jackson K, Spurrier NJ, Gibson J, Whitham J, Williams AS, Dolby R, Kennaway DJ. Behavioral interventions for infant sleep problems: a randomized controlled trial. Pediatrics. 2016;137(6):e20151486. https://doi.org/10.1542/peds.2015-1486. PMID: 27221288.

Sleep in School-Aged Children

7

Janey Dudley, Likhita Shaik, and Shan Luong

Optimal Sleep in School-Aged Children

The National Sleep Foundation outlines specific sleep duration recommendations for children based on their age. For school-aged children, the recommended sleep duration is between 9 and 11 hours per night [2]. Creating an optimal sleeping environment is essential for promoting healthy sleep habits in children. An ideal sleep environment should be dark, quiet, comfortable, safe, and with limited distraction including access to screens. A dark room stimulates the production of melatonin, an endogenous hormone secreted by the pineal gland that promotes sleep and helps to regulate the sleep cycle [3]. Blackout curtains can be beneficial in blocking external light sources. Exposure to screens before bedtime can interfere with sleep quality. It is recommended to limit screen time at least 1 hour before sleep to reduce blue light exposure and its disruption to melatonin production [7]. Loud or distracting noises can make it more difficult for children to fall and stay asleep [4]. White noise machines or soft music can be used to mask disruptive sounds. A comfortable mattress and pillows that provide adequate support are important for a good night's sleep. The temperature of the room should also be kept cool, ideally between 60 and 67 °F (15–19 °C), as a cooler temperature can help with sleep onset and sleep-maintenance [5]. Establishing a regular bedtime routine helps signal to the child that it is time to wind down. This can include activities such as reading or taking a warm bath [6].

Common Barriers to Sleep for Children

Several barriers can hinder children's ability to achieve adequate sleep including increased access to screen time, environmental factors, psychological factors, parental influence, and sleep disorders. Increased use of electronic devices has been linked to delayed bedtimes and reduced sleep duration. The blue light emitted by screens can interfere with the body's natural sleep-wake cycle [7, 8]. Noisy or uncomfortable sleeping environments can disrupt sleep [4]. Parents' sleep habits and attitudes toward sleep can affect their children's sleep patterns. For instance, parents with sleep problems may inadvertently contribute to their children's sleep difficulties [6]. Also, co-sleeping arrangements may lead to sleep disturbances for the child and parents [9]. Anxiety and stress can also contribute to sleep difficulties. Children with attention-deficit/hyperactivity disorder (ADHD) often experience higher rates of sleep problems, which can exacerbate behavioral issues [10]. Other sleep disorders, which will be discussed below, can also significantly impact sleep quality. Research indicates that sleep disorders are prevalent in children, with estimates suggesting that 15–45% of school-aged children experience sleep problems [11].

Sleep Screening Tools for Primary Care Providers

In the context of pediatric primary care, validated screening tools for sleep disorders are essential for early identification and management. Among the various instruments available, the BEARS questionnaire is a concise screening tool that focuses on five key areas: Bedtime problems, Excessive daytime sleepiness, Awakenings during the night, Regularity of sleep schedule, and Snoring. This tool is particularly useful in the busy primary care setting, as it can be completed quickly by parents and provides a comprehensive overview of a child's sleep patterns and potential issues. Electronic screening tools, such as the Children's Sleep Habits Questionnaire (CSHQ), can be answered by parents of children aged 4–10 years. It includes 33 questions that assess multiple dimensions of sleep, including sleep onset, sleep duration, and sleep disturbances, making it a comprehensive tool for identifying potential sleep disorders. Another valuable tool is the 22 question Pediatric Sleep Questionnaire (PSQ), which is designed to identify sleep-related problems in children aged 2 to 18 years. The PSQ includes items that assess sleep-disordered breathing, sleepiness, and behavioral problems associated with sleep disturbances. Collectively, these tools—BEARS, CSHQ, and PSQ—provide primary care providers with a robust framework for screening and addressing sleep disorders in

Fig. 7.1 BEARS questionnaire for primary care providers

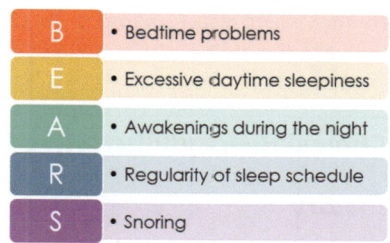

school-aged children, ensuring that potential issues are identified early and managed effectively (Fig. 7.1).

Common Pediatric Sleep Disorders.

Several sleep disorders are more commonly observed in school-aged children: behavioral insomnia, obstructive sleep apnea, parasomnia, restless legs syndrome, periodic limb movement disorder, restless sleep disorder, and circadian rhythm disorders. Many children exhibit resistance to going to bed, often due to a desire to prolong activities or fear of missing out. This can lead to conflicts between parents and children, further complicating bedtime routines [3]. Difficulty falling asleep or staying asleep is prevalent among children. Factors contributing to insomnia can include anxiety, irregular sleep schedules, inadequate parental limit-setting, caffeine, environmental disturbances, and comorbid medical conditions [1, 12]. Obstructive sleep apnea (OSA) is characterized by repeated interruptions in breathing during sleep due to a collapse of the upper airway. Symptoms may include loud snoring, gasping for air, and excessive daytime sleepiness. Studies indicate that OSA is often underdiagnosed in children, despite its significant impact on health and behavior [13]. Parasomnias include abnormal behaviors and perceptions during sleep, such as nightmares, night terrors, sleep talking, sleepwalking. They can be common in children, particularly during periods of stress or change. Children can recall nightmares because they typically occur during REM sleep. Whereas night terrors occur during non-REM sleep and can result in confusion and disorientation upon awakening. While sleepwalking is often benign, it can pose safety risks and may require intervention if occurring frequently [14]. Restless legs syndrome (RLS) is a type of movement disorder characterized by an uncomfortable feeling in the legs and/or arms and can be disruptive to sleep. RLS diagnosis is made clinically, and children should be asked if there is an "urge to move" using age-appropriate prompts. Periodic limb movement disorder requires a diagnosis by polysomnogram that shows ≥5 periodic limb movements per hour and is associated with a sleep-related problem or an impact on daytime functioning. RLS and PLMD are similar but are considered mutually exclusive. More recently, restless sleep disorder (RSD) has been recognized as a pathologic finding in school-aged children characterized by frequent movements during sleep associated with daytime symptoms [15]. RSD also requires PSG that shows ≥5 large muscle group movements per hour. Circadian rhythm sleep-wake disorder in children occurs when there is a misalignment between the child's sleep-wake cycle and the standard time.

Clinical History/Illustrative Case

History

Understanding sleep in school-aged children necessitates a thorough clinical history. For instance, consider a 12-year-old boy, whom we will refer to as "Nick," presented to the pediatric clinic with complaints of excessive daytime sleepiness, difficulty concentrating in school, and a tendency to stay awake late into the night. His mother reported that he often struggles to wake up for school, frequently missing morning classes due to oversleeping. Mom states that, "Nick has to be sleeping at least 12 hours a day considering how late he sleeps in on weekends." The patient reports he has difficulty falling asleep at night and often uses his phone or plays video games to fall asleep. He has no significant past medical history and is not on any medication.

Physical Exam

On physical examination, the patient appeared well-nourished but tired, with dark circles under his eyes. Vital signs were within normal limits. No abnormalities were noted in the cardiovascular, respiratory, or neurological examinations. The patient's body mass index (BMI) was within the normal range. The mental status examination revealed that the patient was alert but exhibited signs of fatigue. His mood was described as "okay," and he denied any feelings of sadness or hopelessness. Cognitive assessment showed mild difficulty with attention and concentration, particularly during the interview. No psychotic symptoms were present, and he denied any suicidal ideation.

Testing or Studies

A screening for depression and anxiety was performed using standardized questionnaires, which returned normal scores. To assess the patient's sleep patterns, a sleep diary was obtained. Nick documented that he would get into bed around 7 PM per his mother's instructions but was unable to fall asleep until around 1–2 AM. He documented that he would wake up at 6 AM on weekdays and around 10–11 AM on weekends. Additionally, an actigraphy study was conducted over a 2-week period, which showed irregular sleep-wake cycles, appropriately correlating with the sleep diary.

Differential Diagnosis

The differential diagnosis for excessive daytime fatigue in this child included: circadian rhythm sleep-wake disorder (CRSWD), insomnia, depression, idiopathic hypersomnia, narcolepsy, and obstructive sleep apnea. Circadian rhythm sleep-wake disorder (CRSWD), more specifically, delayed sleep phase disorder (DSPD) is most likely, given the late sleep onset and wake times. With poor sleep hygiene and late-night electronic device use, true insomnia is less likely, but also on the differential. Depression should always be considered for children with sleep difficulties, especially in combination with fatigue and concentration issues, but ruled out in this case based on mental status exam and screening tools. Per mother's description of patient's sleep habits, idiopathic hypersomnia should be considered; however, Nick's sleep diary and actigraphy do not show excessive sleep duration. Although less likely given the absence of cataplexy or other typical symptoms, narcolepsy should be considered in patients with hypersomnolence. Obstructive sleep apnea is not likely given the normal physical examination and absence of snoring or respiratory symptoms.

Diagnosis and Discussion

The patient was diagnosed with delayed sleep phase disorder (DSPD), a type of circadian rhythm sleep disorder characterized by a significant delay in the timing of sleep onset and wake times compared to societal norms. Recent studies indicate that delayed sleep- wake phase disorder has a range of prevalence up to 16% [16]. This diagnosis was supported by the actigraphy results and the patient's reported sleep patterns, consistent with the criteria outlined in the International Classification of Sleep Disorders. This case illustrates the complexities involved in diagnosing circadian rhythm disorders, particularly when symptoms overlap with other conditions such as hypersomnia, insomnia, and mood disorders. In this case, it was especially important to talk directly to the patient to obtain the most accurate history. The management plan included recommendations for improved sleep hygiene, such as limiting screen time before bed, establishing a consistent sleep schedule, and considering low dose melatonin supplementation to help realign the circadian rhythm [17]. If you are considering melatonin supplements for treatment of DSPD, research supports use of 0.5 mg of fast acting melatonin 4 hours before desired bedtime [18, 21].

Summary

Understanding the importance of sleep in school-aged children is crucial for parents, educators, and healthcare providers. By recognizing the recommended sleep durations, creating optimal sleeping environments, addressing common barriers, and identifying prevalent sleep problems with appropriate screening tools, primary care providers can help promote better sleep health in children. This, in turn, can lead to improved cognitive functioning, emotional regulation, and overall well-being. Sleep disorders in school-aged children are increasingly prevalent and can have long-term consequences on their health and development [19]. The interplay between sleep quality and cognitive function underscores the need for early identification and intervention. Studies have shown that untreated sleep disorders can lead to behavioral issues, academic difficulties, and long-term health problems [20]. Furthermore, the impact of sleep disorders extends beyond the individual child, affecting family dynamics and overall quality of life.

Clinical Pearls

- About one-fourth of all children are reported to have a sleep problem during childhood [1].
- For school-aged children [6–13], the recommended sleep duration is between 9 and 11 hours per night [2].
- BEARS questionnaire is a concise screening tool that focuses on five key areas: Bedtime problems, Excessive daytime sleepiness, Awakenings during the night, Regularity of sleep schedule, and Snoring.
- Studies indicate that OSA is often underdiagnosed in children, despite its significant impact on health and behavior [13].
- Children can recall nightmares because they typically occur during REM sleep. Whereas night terrors occur during non-REM sleep and can result in confusion and disorientation upon awakening.
- If you are considering melatonin supplements for treatment of DSPD, research supports use of 0.5 mg of fast acting melatonin 4 hours before desired bedtime [18, 21].

References

1. Owens J. Insomnia in children and adolescents. J Clin Sleep Med. 2005;1:454–8.
2. Foundation, National Sleep. Back to school sleep tips. National Sleep Foundation Website; 2023.
3. Khazaie H, Zakiei A, Rezaei M, Komasi S, Brand S. Sleep pattern, common bedtime problems, and related factors among first-grade students: Epidemiology and predictors. Clin Epidemiol Glob Health. 2019;7:546–51.
4. Bevan R, Grantham-Hill S, Bowen R, Clayton E, Grice H, Venditti HC, Stickland A, Hill CM. Sleep quality and noise: comparisons between hospital and home settings. Arch Dis Childhood. 2019;104:147–51.
5. Caddick Z, Gregory K, Arsintescu L, Flynn-Evans E. A review of the environmental parameters necessary for an optimal sleep environment. Build Environ. 2018;132:11–20.
6. Etherton H, Blunden S, Hauck Y. Discussion of extinction-based behavioral sleep interventions for young children and reasons why parents may find them difficult. J Clin Sleep Med. 2016;12:1535.
7. Caumo GH, Spritzer D, Carissimi A, Tonon AC. Exposure to electronic devices and sleep quality in adolescents: a matter of type, duration, and timing. Sleep Health. 2020;6:172–8.

8. Hoedlmoser K, Kloesch G, Wiater A, Schabus M. Self-reported sleep patterns, sleep problems, and behavioral problems among school children aged 8-11 years. Somnologie. 2010;14:23–31.
9. Palmer CA, Clementi MA, Meers JM, Alfano CA. Co-sleeping among school-aged anxious and non-anxious children: associations with sleep variability and timing. J Abnorm Child Psychol. 2019;46:1321–32.
10. Yin H, Dong Y, Lin Y, Guangsheng W. Relationship between sleep disorders and attention-deficit-hyperactivity disorder in children. Front Pediatr. 2022;10:919572.
11. Lewien C, Genuneit J, Meigen C, Kiess W, Poulain T. Sleep-related difficulties in healthy children and adolescents. BMC Pediatr. 2021;21:82.
12. Dewald-Kaufmann J, de Bruin E, Michael G. Cognitive behavioral therapy for insomnia (CBT-i) in school-aged children and adolescents. Sleep Med Clin. 2019;14:155–290.
13. Isaiah A, Pereira KD, Das G. Polysomnography and treatment-related outcomes of childhood sleep apnea. Pediatrics. 2019;144:1097.
14. Mason II TBA, Pack AI. Pediatric Parasomnias. Sleep. 2007;30:141–51.
15. DelRosso LM, Mogavero MP, Ferri R, Bruni O. Restless Sleep Disorder (RSD): a new sleep disorder in children. A rapid review. Curr Neurol Neurosci Reports. 2022;22:395–404.
16. Narala B, Ahsan M, Ednick M, Kier C. Delayed sleep wake phase disorder in adolescents: an updated review. Curr Opin Pediatr. 2024;36:124–32.
17. Sun S-Y, Chen G-H. Treatment of circadian rhythm sleep–wake disorders. Curr Neuropharmacol. 2022;20:1022.
18. Sletten T, Magee M, Murray J, Gordon C, Lovato N, Kennaway D, Gwini S, Bartlett D, Lockley S, Lack L, Grunstein R, Rajaratnam S. Efficacy of melatonin with behavioural sleep-wake scheduling for delayed sleep-wake phase disorder: a double-blind, randomised clinical trial. PLOS Med. 2018;15
19. Sundell AL, Angelhoff C. Sleep and its relation to health-related quality of life in 3–10-year-old children. BMC Public Health. 2021;21:1043.
20. Kelly RJ, El-Sheikh M. Reciprocal relations between children's sleep and their adjustment over time. Dev Psychol. 2014;50:1137–47.
21. Mantle D, Smits M, Boss M, Miedema I, van Geijlswijk I. Efficacy and safety of supplemental melatonin for delayed sleep–wake phase disorder in children: an overview. Sleep Med. 2020;2:100022.

Sleep in Adolescents

8

Crystal Cassimere and Michelle Caraballo

Physiology of Sleep in Adolescents

Sleep in adolescents is influenced by multiple factors, including shifts in circadian rhythm, discrepancy between sleep needs versus actual sleep patterns, and psychosocial and behavioral factors (Fig. 8.1) [1, 2]. A natural shift in circadian rhythm occurs during adolescence, often referred to as a phase delay, whereby teenagers have a natural tendency to fall asleep later and wake up later compared to younger children [1, 3]. This shift is driven by biological changes, including a delay in the onset of melatonin secretion. In adolescents, melatonin release typically occurs between 9:00 p.m. and 11:30 p.m., resulting in delayed natural sleep onset [1, 3]. Morning melatonin clearance also tends to be delayed, making it harder for adolescents to wake up early [3]. The phase delay aligns poorly with early school start times, contributing to chronic sleep deprivation in this age group [1]. Adolescents often struggle with "social jetlag," where their natural sleep-wake patterns are misaligned with societal demands [4].

Sleep architecture refers to the organization of different sleep stages throughout the night. Adolescents, like adults, cycle through the stages of sleep approximately every 90 minutes [6]. Non-rapid eye movement (NREM) sleep includes stages 1 through 3, with stage 3 representing deep slow-wave sleep. Slow-wave sleep is critical for restorative functions, memory consolidation, and growth [6]. Adolescents experience a reduction in the proportion of slow-wave sleep compared to younger

---

C. Cassimere
Department of Family Medicine, University of Texas Southwestern Medical Center, Dallas, TX, USA
e-mail: Crystal.Cassimere@UTSouthwestern.edu

M. Caraballo (✉)
Department of Pediatrics, Division of Pediatric Pulmonology and Sleep Medicine, University of Texas Southwestern Medical Center, Dallas, TX, USA
e-mail: Michelle.Caraballo@UTSouthwestern.edu

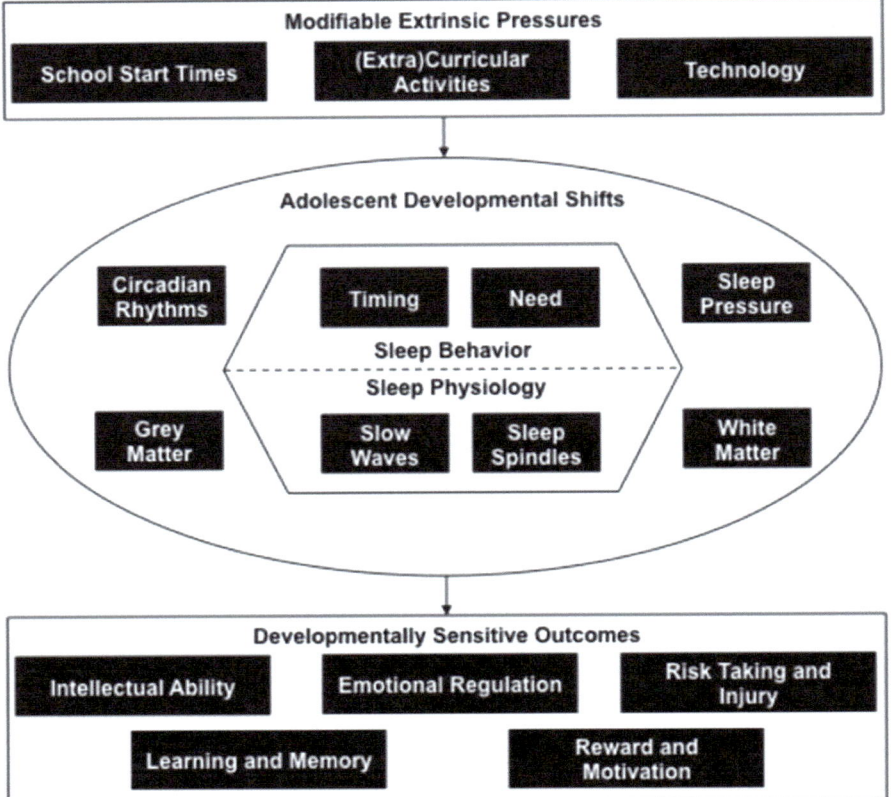

Fig. 8.1 Several complex factors affect sleep during adolescence, and developmental changes in sleep behavior and physiology coincide with brain maturation (center). External pressures like school schedules and technology use can impact sleep (top), ultimately affecting various aspects of behavior and cognition during this sensitive period (bottom). (Reprinted from: "Sleep in adolescence: Physiology, cognition and mental health" by Tarokh et al. [5])

children (Fig. 8.2). As adolescents grow, the distribution of sleep stages evolves, with a decrease in slow-wave sleep and an increase in lighter sleep stages. Rapid eye movement (REM) sleep accounts for 20–25% of total sleep and is essential for emotional regulation and cognitive processing [6]. Adolescents experience a small increase in REM proportions compared to younger children, but the timing and duration of REM periods may shift due to the phase delay.

These changes in sleep stage distribution contribute to the vulnerability of adolescent sleep to disruption from environmental or behavioral factors. Adolescents typically need 8–10 hours of sleep per night for optimal health, cognitive functioning, and emotional well-being. The CDC estimates that 7 in 10 adolescents do not get enough sleep on school nights owing to multiple factors, including delayed circadian rhythm, academic pressures, social activities, and increased electronic use

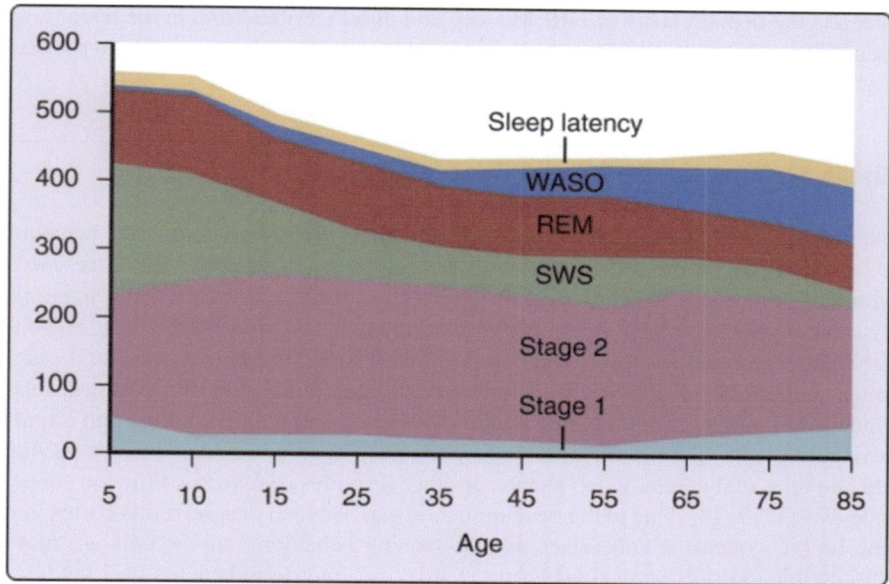

Fig. 8.2 This graph shows age-related trends for sleep stage distribution, wake after sleep onset (WASO), and sleep latency (in minutes). The proportion of slow-wave sleep declines throughout adolescence with associated increase in stage 2 sleep. There is also a modest increase in percentage of REM sleep from childhood to the end of adolescence. (Reprinted from Ohayon et al. [10])

[7, 8]. Insufficient sleep is linked to mood disturbances, decreased academic performance, and increased risk-taking behaviors [2, 4, 7, 9]. Long-term sleep deprivation can also contribute to metabolic disturbances, obesity, and impaired immune function [1, 9]. Many adolescents try to "catch up" on sleep during weekends, leading to irregular sleep patterns and worsening the misalignment between their biological clock and weekday schedules [1, 4]. This cycle of sleep debt exacerbates the phase delay.

Developmental and Biological Changes in Adolescent Sleep

The onset of puberty triggers significant hormonal changes that impact sleep patterns. These changes are driven primarily by the activation of the hypothalamic-pituitary-gonadal (HPG) axis, which regulates reproductive hormones like estrogen and testosterone. These hormones influence both circadian rhythms and sleep architecture [12]. As discussed previously, puberty is associated with a delay in melatonin release, shifting the natural timing of sleep onset and wakefulness. The circadian rhythm shifts later in the evening, making it harder for adolescents to fall asleep early [1, 3, 5, 11]. Slow-wave sleep is closely linked to the secretion of growth hormone, which is critical during the adolescent growth spurt [12]. Since slow-wave sleep is the deepest stage of NREM sleep and plays a critical role in the restorative benefit from sleep, the decreasing proportion of slow-wave sleep throughout puberty can impact overall sleep quality and recovery [12].

Brain Maturation and Sleep Regulation

Adolescent brain maturation significantly impacts sleep regulation and behavior due to changes in the prefrontal cortex and other brain regions. Various research supports how these developmental processes influence adolescent sleep patterns, risk-taking behavior, and emotional regulation [13]. The prefrontal cortex, responsible for executive functions such as decisionmaking, impulse control, and planning, continues to mature well into early adulthood [13, 14]. Adolescents' underdeveloped prefrontal cortex makes them more prone to risk-taking and impulsive behavior [14]. This can manifest in poor sleep hygiene choices, such as staying up late on social media, video games, or other activities despite knowing the consequences [4, 13, 14]. Due to the developmental gap between the prefrontal cortex and the limbic system, which drives reward-seeking behaviors, adolescents are more susceptible to prioritizing short-term rewards over long-term benefits [4, 14, 15].

Synaptic pruning is a process where the brain eliminates less-used neural connections while strengthening essential ones, refining the brain's neural networks [16].

This process peaks during adolescence and is crucial for cognitive development, including learning and memory. Slow-wave sleep is especially important for synaptic pruning [16, 17]. Research shows that adequate slow-wave sleep contributes to memory consolidation and cognitive efficiency, while sleep deprivation can disrupt these processes, leading to impaired learning and cognitive function [15–17].

The sleep-wake cycle undergoes a shift in adolescence due to changes in homeostatic sleep drive and circadian rhythms [1, 2]. Adolescents experience a delay in their circadian phase, often leading to later sleep onset and wake times. This delay is partially due to the reduced build-up of sleep pressure in the evening [1, 2]. Consequently, adolescents can stay awake longer, even if they are accumulating sleep debt, leading to chronic sleep deprivation. This shift often conflicts with early school start times, resulting in sleep deprivation and daytime fatigue [1, 2].

Sleep and emotional regulation are closely interconnected, with significant implications for adolescent mental health. The prefrontal cortex's immaturity, combined with hormonal changes and psychosocial stressors, makes adolescents particularly vulnerable to mood swings, anxiety, and depression [14, 15]. Sleep deprivation impairs the ability to manage emotions, heightening irritability and increasing the risk of mental health issues, leading to a vicious cycle where poor emotional regulation further disturbs sleep [15].

Common Sleep Disorders in Adolescents

Delayed Sleep-Wake Phase Disorder (DSWPD)

Delayed Sleep-Wake Phase Disorder (DSWPD) is a circadian rhythm disorder estimated to affect 7–16% of adolescents and young adults, characterized by a significant delay in the timing of sleep onset and wake-up times, usually by more than 2 hours compared to societal norms (Fig. 8.3) [18]. Adolescents with DSWPD often struggle to fall asleep at the desired or necessary bedtime, despite efforts at consistent sleep schedule and healthy sleep hygiene. They may be unable to fall asleep until late at night (e.g., 1–4 a.m.) and have difficulty waking up early, leading to daytime sleepiness and impaired functioning. DSWPD is more common in this age group due to the natural circadian phase delay during puberty, and these physiologic changes are compounded by lifestyle factors like late-night screen use and irregular sleep schedules [1, 18, 19].

Diagnosis is through self-reported symptoms and sleep schedules corroborated by actigraphy, a small wrist-worn device that tracks sleep and wake schedules over the course of 7–14 days, including both school/workdays and days off [18, 19]. Management consists of behavioral strategies, chronotherapy, and strategic timing of light therapy and melatonin supplementation to help shift the circadian rhythm [18–20]. Chronotherapy involves gradually shifting sleep and wake times earlier each day to reset the circadian rhythm [18–20]. Exposure to bright light in the morning can help advance the sleep-wake phase by signaling the brain to shift the circadian clock earlier [18–20]. Low doses of melatonin strategically administered in the afternoon or early evening can help shift the circadian rhythm earlier (Fig. 8.4) [18–20]. Maintaining consistent sleep schedules and strict sleep hygiene, including limiting evening exposure to electronics, bright lights, and stimulating activities, are crucial components of treatment.

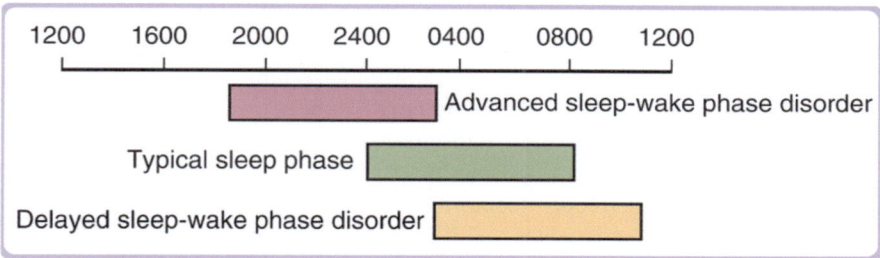

Fig. 8.3 Circadian Disorders of the Sleep-Wake Cycle. Patients with delayed sleep-wake phase disorder often have trouble falling asleep before 2 a.m. and struggle to wake in the morning, while those with advanced sleep-wake phase disorder feel evening sleepiness and wake early. Reprinted from Harsanyi et al., Published 2021, Elsevier [18]

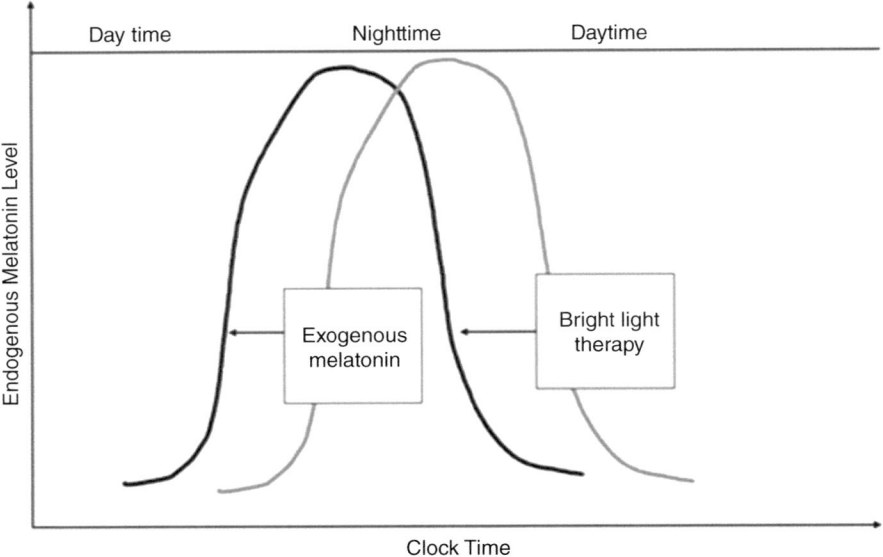

Fig. 8.4 This image depicts the changes in endogenous melatonin levels throughout the day. People with normal sleep patterns (black line) experience an increase in melatonin about 2 hours before sleep onset, while those with delayed sleep phases (gray line) have a later rise. Taking melatonin in the evening and using bright light in the morning can help shift sleep timing earlier. (Reprinted from Krishna et al. [20])

Insomnia

Insomnia is characterized by frequent or persistent difficulty initiating or maintaining sleep, resulting in inadequate sleep quality or duration, despite adequate sleep opportunity (i.e., time available for sleep) and circumstances (i.e., darkness, quietness, comfort) [18, 20]. The patient or parent must also describe at least one symptom associated with sleep disturbance, such as concern or distress about lack of sleep, daytime sleepiness, behavioral problems, or impaired social or academic performance [18, 20]. Academic pressures, social dynamics, and family issues are common triggers for insomnia in this age group [21, 22]. Technology use also exacerbates insomnia [23]. Excessive screen time, especially before bed, disrupts sleep due to overstimulation and blue light exposure, which inhibits the natural melatonin release that induces sleepiness [23]. One large study of insomnia in adolescents showed that the median age of onset of insomnia was 11, and over half of those pediatric patients with insomnia had a comorbid psychiatric disorder [24]. Anxiety and depression are the most frequent comorbid conditions, and the bidirectional relationship between sleep and mental health can exacerbate both conditions [20, 21].

Cognitive Behavioral Therapy for Insomnia (CBT-I) is the first-line treatment for insomnia in adolescents, focusing on addressing negative thoughts, behaviors, and habits that contribute to sleep difficulties [25]. Other non-pharmacologic treatment modalities include teaching adolescents about healthy sleep practices, such as maintaining a regular sleep schedule, avoiding caffeine and naps, and creating a calming bedtime routine to develop good sleep hygiene [20]. Pharmacological treatment is generally considered a secondary option and should be used cautiously, with melatonin being the most commonly recommended [20, 25].

Obstructive Sleep Apnea

Obstructive sleep apnea (OSA) is a sleep-related breathing disorder characterized by episodes of partial or complete airway obstruction during sleep, leading to disrupted sleep and reduced oxygen saturation [1, 20, 26]. Symptoms include habitual snoring, witnessed apneas, gasping for air, and impaired daytime functioning such as daytime sleepiness, behavior problems, or poor school performance [1, 20, 26]. Risk factors include obesity, adenotonsillar hypertrophy, and craniofacial abnormalities [1, 20, 26]. Polysomnography (sleep study) is the gold standard for diagnosing OSA [1, 20, 26]. While screening questionnaires and clinical evaluation help identify at-risk adolescents, the sensitivity and specificity of history and physical exam alone as screening tools to diagnose OSA are poor. Weight management through lifestyle modifications can improve OSA symptoms in obese adolescents but is typically not recommended as the only treatment [1, 20, 26]. Adenotonsillectomy is first-line therapy for OSA in pediatric patients [1, 20, 26]. Continuous Positive Airway Pressure (CPAP) is the mainstay of therapy in adolescents with residual OSA after adenotonsillectomy or in patients who are not good candidates for adenotonsillectomy due to obesity, small tonsils/adenoids, or other comorbid medical conditions making them higher risk for surgery [1, 20, 26]. Oral appliances or other orthodontic interventions may also be considered in certain cases to adjust airway anatomy [1, 20, 26].

Parasomnias

Parasomnias are physical or verbal behaviors occurring during sleep, including sleepwalking (somnambulism), sleep talking, night terrors, and confusional arousals [20]. They typically occur during NREM stage 3 sleep (slow-wave sleep) and are more common in younger children but can persist or reemerge in adolescence [20]. Sleepwalking presents as episodes of walking or performing other complex behaviors while in a state of partial arousal. Parasomnias may be triggered in susceptible individuals by sleep deprivation, irregular sleep schedules, illness, or stress [20]. Night terrors, which are uncommon in adolescence, are sudden episodes of intense fear, screaming, and autonomic arousal (e.g., rapid heart rate). The individual is usually difficult to awaken and has no memory of the event upon waking.

Parasomnias can be very frightening to parents and caregivers but are generally benign and not associated with any underlying neurological disorder [20]. Ensuring a safe environment is critical (e.g., locking doors and removing obstacles) to prevent injuries during episodes. Timed wakings, consisting of gently waking the patient 15–30 minutes before the usual time of a parasomnia episode, can reduce their occurrence by disrupting the sleep cycle [20]. Maintaining good sleep hygiene with a consistent sleep schedule and adequate sleep opportunity can decrease the frequency of parasomnias. Reducing stress and managing other potential triggers, such as medication side effects, are also important aspects of management [20].

Psychosocial and Behavioral Factors

Psychosocial and behavioral factors play a critical role in the sleep patterns of adolescents, often contributing to various sleep disturbances. The demand to excel in school, participate in extracurricular activities, and maintain a social life can lead to overextended teens with inadequate sleep opportunity and chronic stress. This stress frequently manifests as disturbed mood, difficulty falling asleep, poor sleep quality, and insufficient sleep duration. These pressures are further exacerbated by early school start times, which are often misaligned with the natural sleep-wake cycles of teenagers [21]. Research consistently supports the idea that later school start times can improve sleep duration, mood, and academic performance [1, 2, 5, 7, 21]. The American Academy of Pediatrics (AAP) recommends that middle and high schools start no earlier than 8:30 a.m. for optimal adolescent sleep, health, and learning. Despite this evidence, many schools have earlier start times, worsening sleep deprivation and its associated consequences [21].

Technology and screen time also have a significant impact on adolescent sleep. The widespread use of electronic devices before and even after bedtime has been shown to disrupt sleep due to the emission of blue light, which suppresses melatonin release, and frequently noise, which also disrupts an optimal sleep environment [23]. This delay in melatonin release interferes with sleep onset, leading to prolonged sleep latency and reduced overall sleep quality. Additionally, the content adolescents engage with on screens can heighten arousal and make it harder to wind down, further delaying sleep.

Peer relationships and social connectivity further complicate adolescent sleep issues.

Adolescents are highly influenced by their peers, which often results in late-night interactions on social media or messaging platforms. The pressure to stay connected and the fear of missing out (FOMO) can lead to delayed bedtimes and disrupted sleep. Moreover, the impact of social media extends beyond delayed bedtimes; it also correlates with increased anxiety and sleep disturbances, creating a vicious cycle where poor sleep exacerbates stress and vice versa [15]. The bidirectional relationship between sleep disturbances and mental health conditions like depression and anxiety is another critical consideration in adolescents. Poor sleep can exacerbate symptoms of depression, while depression itself often leads to insomnia and irregular sleep patterns. Similarly, anxiety disorders are closely tied to sleep problems, with worrying thoughts making it difficult to initiate sleep. This cyclical relationship means that each condition can worsen the other, making it challenging to break the pattern [15].

Asking patients about sleep concerns and changes can be a helpful tool for identifying early signs of mental health issues. Significant alterations in sleep patterns, such as prolonged insomnia, excessive sleepiness, or erratic sleep schedules, can be early indicators of conditions like depression or anxiety [20]. Incorporating sleep assessments into mental health evaluations and even routine primary care checkups provides a more comprehensive understanding of the challenges adolescents face. This holistic approach enables earlier intervention and more targeted treatment strategies, addressing both sleep disturbances and any underlying mental health concerns contributing to them.

Cultural and Socioeconomic Considerations

Cultural and socioeconomic factors play a significant role in shaping adolescent sleep behaviors and patterns, influencing both the quantity and quality of sleep [27]. Cultural norms regarding bedtime routines, co-sleeping practices, and attitudes toward napping can lead to notable differences in sleep behaviors across various communities. For example, in some cultures, later bedtimes are the norm, which often results in shorter sleep duration during school nights. Family dynamics and cultural practices, such as shared sleeping arrangements or specific routines, can either support or hinder healthy sleep patterns. These culturally-rooted behaviors and values highlight how sleep is not just a biological process but also a practice deeply intertwined with social and cultural contexts.

In addition to cultural influences, socioeconomic status (SES) plays a crucial role in sleep quality and access to resources that promote healthy sleep. Adolescents from lower socioeconomic backgrounds often face environmental challenges such as overcrowded living spaces, excessive noise, and unsafe neighborhoods, all of which can disrupt sleep. The stress associated with financial instability, food insecurity, and other SES-related hardships further exacerbates sleep difficulties [27]. Chronic stress in these environments can lead to sleep disturbances, which in turn negatively affect general health and well-being. Moreover, adolescents in lower SES settings may have limited access to healthcare resources that could address sleep-related issues, making it more challenging to receive proper treatment and support.

Understanding the interplay between cultural and socioeconomic factors is essential in addressing sleep disparities among adolescents. These influences highlight the need for culturally-sensitive and context-specific approaches to improving sleep health, ensuring that interventions are tailored to meet the unique needs and challenges faced by diverse populations. By recognizing the broader social determinants that impact sleep, strategies can be developed to promote better sleep practices that take into account the cultural and economic realities of adolescents' lives.

Impact of Sleep on Adolescent Health and Well-Being

Sleep is a critical factor in the health and well-being of adolescents, affecting everything from academic performance to physical health and behavioral choices. The relationship between sleep and cognitive functioning is particularly significant during adolescence, a period when learning and memory are central to daily activities. Sleep plays an essential role in memory consolidation and cognitive performance. When adolescents experience sleep disturbance or deprivation, their ability to pay attention, process information, and retain new knowledge is severely compromised [6, 15–17]. This lack of cognitive efficiency often translates into poor academic performance, with reduced classroom engagement and lower achievement.

In addition to cognitive impacts, sleep is deeply connected to physical health. Adolescents who do not get adequate sleep face a heightened risk of obesity and metabolic disorders. Sleep deprivation disrupts the balance of appetite-regulating hormones like ghrelin and leptin, leading to increased cravings for high-calorie foods [12]. This imbalance, coupled with lower energy levels that reduce physical activity, creates a perfect storm for weight gain. Furthermore, poor sleep weakens the immune system, making adolescents more susceptible to infections and hindering recovery from illnesses [2]. The cumulative effect of inadequate sleep on metabolic health and immunity highlights how critical proper rest is for overall well-being.

Overall, the impact of sleep on adolescent health and well-being is profound and multifaceted. From academic success to physical health and behavioral choices, adequate sleep is a cornerstone of healthy development during these formative years. Addressing sleep issues in adolescents is not just a matter of improving rest, it is an essential step toward fostering cognitive, physical, and emotional resilience in young people as they navigate the challenges of adolescence.

Clinical Pearls

- Adolescent sleep is influenced by multiple developmental, biological, and behavioral factors and changes that accompany pubertal development.
- Many adolescents experience a circadian phase delay, which shifts intrinsic sleep and wake times later and can disrupt ability to participate and function well in daytime activities including school, work, and social activities.
- Slow-wave sleep decreases during adolescence, which may have negative effects on memory and emotional regulation.
- Most adolescents do not get adequate sleep on a regular basis, leading to impaired school performance, mood disturbances, and risk-taking behaviors.
- Evaluation and management of sleep disturbances are crucial to adolescent health and well-being.

References

1. Sheldon SH, Ferber R, Kryger MH, Gozal D, editors. Principles and practice of pediatric sleep medicine. 2nd ed. Elsevier; 2014.
2. Carskadon MA. Sleep in adolescents: the perfect storm. Pediatrics. 2011;128(3):613–9.

3. National Research Council (US) & Institute of Medicine (US) Forum on Adolescence, Graham, M. G. (Ed.). Sleep needs, patterns, and difficulties of adolescents: summary of a workshop. National Academies Press (US); 2000. https://www.ncbi.nlm.nih.gov/books/NBK222804/.
4. Tonetti L, Andreose A, Bacaro V, Grimaldi M, Natale V, Crocetti E. Different effects of social jetlag and weekend catch-up sleep on well-being of adolescents according to the actual sleep duration. Int J Environ Res Public Health. 2023;20(1):574. https://doi.org/10.3390/ijerph20010574.
5. Tarokh L, Saletin JM, Carskadon MA. Sleep in adolescence: Physiology, cognition and mental health. Neurosci Biobehav Rev. 2016;70:182–8. https://doi.org/10.1016/j.neubiorev.2016.08.008. Epub 2016 Aug 13. PMID: 27531236; PMCID: PMC5074885.
6. Carskadon MA, Dement WC. Monitoring and staging human sleep. In: Kryger MH, Roth T, Dement WC, editors. Principles and practice of sleep medicine. 5th ed. St. Louis: Elsevier Saunders; 2011. p. 16–26.
7. Wheaton AG, Jones SE, Cooper AC, Croft JB. Short sleep duration among middle school and high school students — United States, 2015. MMWR Morb Mortal Wkly Rep. 2018;67:85–90.
8. Hena M, Garmy P. Social jetlag and its association with screen time and nighttime texting among adolescents in Sweden: a cross-sectional study. Front Neurosci. 2020;14:122. https://doi.org/10.3389/fnins.2020.00122. PMID: 32132896; PMCID: PMC7040091.
9. Beebe DW. Cognitive, behavioral, and functional consequences of inadequate sleep in children and adolescents. Pediatr Clin N Am. 2011;58(3):649–65.
10. Ohayon MM, Carskadon MA, Guilleminault C, Vitiello MV. Meta-analysis of quantitative sleep parameters from childhood to old age in healthy individuals: developing normative sleep values across the human lifespan. Sleep. 2004;27(7):1255–73. https://doi.org/10.1093/sleep/27.7.1255.
11. Hagenauer MH, Perryman JI, Lee TM, Carskadon MA. Adolescent changes in the homeostatic and circadian regulation of sleep. Dev Neurosci. 2009;31(4):276–84.
12. Lucien JN, Ortega MT, Shaw ND. Sleep and puberty. Curr Opin Endocr Metab Res. 2021;17:1–7. https://doi.org/10.1016/j.coemr.2020.09.009. Epub 2020 Oct 9. PMID: 35005296; PMCID: PMC8730357.
13. Casey BJ, Jones RM, Hare TA. The adolescent brain. Ann N Y Acad Sci. 2008;1124:111–26. https://doi.org/10.1196/annals.1440.010. PMID: 18400927; PMCID: PMC2475802.
14. Johnson SB, Blum RW, Giedd JN. Adolescent maturity and the brain: the promise and pitfalls of neuroscience research in adolescent health policy. J Adolesc Health. 2009;45(3):216–21. https://doi.org/10.1016/j.jadohealth.2009.05.016. PMID: 19699416; PMCID: PMC2892678.
15. Goldstein AN, Walker MP. The role of sleep in emotional brain function. Annu Rev Clin Psychol. 2014;10:679–708. https://doi.org/10.1146/annurev-clinpsy-032813-153716. Epub 2014 Jan 31. PMID: 24499013; PMCID: PMC4286245.
16. Tuan L, Lee L. Microglia-mediated synaptic pruning is impaired in sleep-deprived adolescent mice. Neurobiol Dis. 2019;130:104517. https://doi.org/10.1016/j.nbd.2019.104517.
17. Tononi G, Cirelli C. Sleep and the price of plasticity: from synaptic and cellular homeostasis to memory consolidation and integration. Neuron. 2014;81(1):12–34.
18. Principles and Practice of Sleep Medicine Harsanyi K, et al. Chapter 70—Epidemiology of sleep medicine. In: Kryger MH, Roth T, Goldstein CA, editors. Principles and practice of sleep medicine. 7t ed. Elsevier; 2021. p. 689.
19. Crowley SJ, Acebo C, Carskadon MA. Sleep, circadian rhythms, and delayed phase in adolescence. Sleep Med. 2007;8(6):602–12. https://doi.org/10.1016/j.sleep.2006.12.002.
20. Krishna J, Kalra M, McQuillan ME. Sleep disorders in childhood. Pediatr Rev. 2023;44(4):189–202. https://doi.org/10.1542/pir.2022-005521.
21. Au R, Carskadon M, Millman R, Wolfson A, Braverman PK, Adelman WP, Breuner CC, Levine DA, Marcell AV, Murray PJ, O'Brien RF, Devore CD, Allison M, Ancona R, Barnett FSE, Gunther R, Holmes B, Lamont JH, Minier M, et al. School start times for adolescents. Pediatrics. 2014;134(3):642–9. https://doi.org/10.1542/peds.2014-1697.

22. Haspolat NK, Ağirkan M. When parents press for achievement: the relationship between academic stress, insomnia, adolescent-parent relationships, and life satisfaction. J Child Fam Stud. 2024;33:3486. https://doi.org/10.1007/s10826-024-02921-z.
23. Hale L, Kirschen GW, LeBourgeois MK, Gradisar M, Garrison MM, Montgomery-Downs H, Kirschen H, McHale SM, Chang AM, Buxton OM. Youth screen media habits and sleep: sleep-friendly screen behavior recommendations for clinicians, educators, and parents. Child Adolesc Psychiatr Clin N Am. 2018;27(2):229–45. https://doi.org/10.1016/j.chc.2017.11.014. PMID: 29502749; PMCID: PMC5839336.
24. Johnson EO, Roth T, Schultz L, Breslau N. Epidemiology of DSM-IV insomnia in adolescence: lifetime prevalence, chronicity, and an emergent gender difference. Pediatrics. 2006;117(2):e247–56. https://doi.org/10.1542/peds.2004-2629.
25. Dewald-Kaufmann J, De Bruin E, Michael G. Cognitive behavioral therapy for insomnia (CBT-I) in school-aged children and adolescents. Sleep Med Clin. 2019;14(2):155–65. https://doi.org/10.1016/j.jsmc.2019.02.002.
26. Marcus CL, Brooks LJ, Draper KA, Gozal D, Halbower AC, Jones J, Schechter MS, Sheldon SH, Spruyt K, Ward SD, Lehmann C, Shiffman RN. Diagnosis and management of childhood obstructive sleep apnea syndrome. Pediatrics. 2012;130(3):576–84. https://doi.org/10.1542/peds.2012-167.
27. Philbrook LE, Saini EK, Fuller-Rowell TE, Buckhalt JA, El-Sheikh M. Socioeconomic status and sleep in adolescence: the role of family chaos. J Fam Psychol. 2020;34(5):577–86. https://doi.org/10.1037/fam0000636. Epub 2020 Feb 3. PMID: 32011158; PMCID: PMC7374040.

Part III
Common Sleep Disorders

Obstructive Sleep Apnea

9

Aarti Shakkottai and Manahil Firdaus

Introduction

Obstructive sleep apnea affects up to 13% of children in the United States [1].

It is characterized by recurrent episodes of partial (hypopnea) or complete (apnea) airway collapse during sleep that is frequently accompanied by a drop in oxygen saturation (hypoxemia) and/or an arousal from sleep. One of the hallmarks of OSA is the presence of respiratory effort during these episodes of airway obstruction/collapse. This helps differentiate it from central sleep apnea (CSA), which is also characterized by recurrent apneas and hypopneas but without any respiratory effort. Sleep-related hypoventilation, defined as a carbon dioxide value that is greater than 50 mmHg for more than 25% of the total sleep time, is often an accompanying feature of OSA. When it is accompanied by snoring and paradoxical chest and abdominal wall movement, then it is called obstructive hypoventilation, an entity that is analogous to OSA in children (ICSD-3-TR) [2]. Importantly, OSA may present with behavioral symptoms that mimic neurodevelopmental disorders, particularly attention-deficit/hyperactivity disorder (ADHD). Daytime manifestations of OSA in children can include hyperactivity, inattention, impulsivity, and poor academic performance—symptoms that overlap significantly with ADHD. As rates of ADHD diagnoses continue to rise, it is critical that primary care providers rule out sleep-disordered breathing, including OSA, before initiating a diagnosis or treatment plan for ADHD. Identifying and treating OSA may not only improve sleep and overall health but also lead to significant improvements in behavior and cognitive function, potentially reducing the need for pharmacologic intervention.

A. Shakkottai (✉)
Pediatric Pulmonology and Sleep Medicine, University of Texas Southwestern Medical Center, Dallas, TX, USA
e-mail: Aarti.Shakkottai@UTSouthwestern.edu

M. Firdaus
University of Texas at Dallas, Richardson, TX, USA

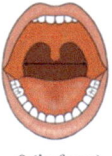 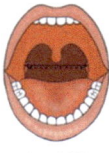 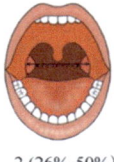 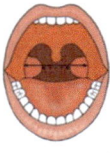 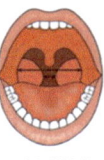

0 (in fossa)   1 (≤25%)   2 (26%-50%)   3 (51%-75%)   4 (>75%)

Fig. 9.1 Brodsky grading scale of tonsil size. Grade 0 (tonsils in the fossa), grade 1 (tonsils outside of the fossa and occupy ≤25% of the oropharyngeal width), grade 2 (tonsils occupy 26–50% of the oropharyngeal width), grade 3 (tonsils occupy 51–75% of the oropharyngeal width), and grade 4 (tonsils occupy > 75% of the oropharyngeal width). (From paper by Lv et al. BMC Oral Health. 2023 Nov 7, 23(1):836)

Clinical Presentation and Exam Findings

While the presentation of pediatric OSA can be heterogeneous, common symptoms associated with OSA include snoring, mouth breathing, nocturnal diaphoresis, restless and fragmented sleep, nocturnal enuresis, sleeping in unusual positions including with the neck hyperextended, waking up with a dry mouth, and daytime sleepiness. Among young children, it is more common to see inattentiveness, irritability, or hyperactivity rather than hypersomnolence [3, 4].

Physical exam findings of OSA include adenoid facies (sunken eyes, open mouth, pinched nose, hypoplastic maxilla, high-arched palate, crowded teeth, retrognathia, see picture), nasal obstruction (septal deviation, hypertrophied nasal turbinates), midface hypoplasia, retrognathia, micrognathia, palatal abnormalities including a low-lying palate or narrow high-arched palate, malocclusion, large broad tongue with scalloping, abnormal voice quality, and enlarged tonsils (graded using the Brodsky scoring criteria, see Fig. 9.1) [5]. Children with OSA can be overweight/obese. Poor weight gain or failure to thrive can also be a manifestation of OSA in children [6]. Children with OSA have been shown to be at increased risk for systemic and pulmonary hypertension and cardiac remodeling including diastolic dysfunction, which may also be evident on the physical exam [7].

Risk Factors

There are several risk factors for OSA in children, with obesity being among the most prominent. According to a population study of 400 children between ages 2 and 18, obesity was found to be the most significant risk factor for OSA based on an odds ratio of 4.69 [8]. Per every unit increase in BMI beyond the 50th percentile, the risk for OSA increased by 10% [9]. The association between obesity and OSA is hypothesized to be a consequence of excess adipose tissue deposition in the upper airway including in areas of the soft palate and tongue, increased mechanical load on the chest wall, and lymphoid hyperplasia contributing to adenoid and tonsil volume [10–14]. Adenoid and tonsillar hypertrophy is another important risk factor for OSA among children [15]. Young children (ages 2–8 years) tend to have enlargement of their adenoids and tonsils relative to the rest of their airway and are consequently at greater risk for OSA [16]. Frequency of OSA tends to be similar between prepubertal boys and girls. Post-puberty, there is an increased risk of OSA in boys as compared to girls [17, 18]. Children with certain genetic conditions, such as Down Syndrome, craniofacial anomalies including Pierre-Robin sequence and Apert syndrome, and neuromuscular disorders such as Duchenne muscular dystrophy, are also at increased risk for OSA [5]. The presence of asthma is also an important risk factor for OSA, with studies suggesting a four-times increased risk for sleep-disordered breathing among asthmatics as compared to those without this comorbidity [19]. Prematurity is another risk factor for OSA [20]. Second-hand smoke exposure (SHS), both prenatally and during childhood, can contribute to OSA severity. Odds of having a higher obstructive apnea index was 1.48 times greater among children with OSA who experienced SHS exposure as compared to those who did not [21]. Historically, certain races such as African Americans, Asians, and Native Americans have been suggested to be at increased risk for OSA. Whether this increased risk is genetically mediated versus secondary to socioeconomic inequities needs to be further explored.

Diagnosis

The gold standard for the diagnosis of OSA in children is an overnight in-laboratory polysomnogram, or sleep study. This study synthesizes information from electroencephalography (EEG), electrooculography (EOG), electromyography (EMG), electrocardiography (ECG), pulse oximetry, airflow, and respiratory effort channels to diagnose OSA. In children, an apnea-hypopnea index (AHI) greater than 1 per hour of sleep is considered diagnostic for OSA. An AHI ≥5 per hour of sleep indicates moderate OSA and an AHI ≥10 per hour of sleep is considered severe OSA. Home sleep apnea testing (HSAT), which uses a combination of airflow, respiratory effort, pulse oximetry, and ECG to identify OSA, has become the standard of care for screening adults for OSA. Although there is a growing body of data supporting the use of HSATs in children, these devices are still not considered standard of care in children [22, 23]. Overnight pulse oximetry has been used to screen for OSA in children [24]. There are several validated screening questionnaires, such as the Pediatric Sleep-Questionnaire—Sleep-Related Breathing Disorders (PSQ-SRBD) Scale and Obstructive Sleep Apnea–18 that have also been used to identify OSA symptoms in children [25]. The OSA-18 is a health-related quality of life tool that looks at the impact of OSA symptoms on quality of life including physical, emotional, and social functioning [26]. Radiographs of the upper airway/nasopharynx have been used to look for adenoid and tonsillar hypertrophy [5]. Computerized tomography (CT) scans and magnetic resonance imaging (MRI) of the upper airway have also been used to identify sources of upper airway obstruction in children with OSA. Lastly, drug-induced sleep endoscopy (DISE) has been frequently used in children with persistent OSA despite adenotonsillectomy to look for other sites of airway obstruction [5].

Treatment

Adenotonsillectomy is considered first-line of treatment in children with OSA who have evidence of adenoid and tonsillar hypertrophy [27, 28]. It is curative in up to 80% of cases of OSA [29]. It has been shown to normalize the AHI and bring about improvements in quality of life, behavior, and mood [30, 31]. A preoperative polysomnogram is recommended in children <2 years of age, those with Down syndrome, craniofacial anomalies, neuromuscular disease, or obesity, as these are also individuals who are at increased risk for residual OSA after adenotonsillectomy [28]. Alternatives to adenotonsillectomy, including partial tonsillectomy or tonsillotomy, have gained traction in recent years due to reduced postoperative morbidity and quicker recovery time. However, higher rates of tonsillar regrowth undermine the viability of these procedures as first-line treatment [32]. Other surgical options in children with residual OSA include nasal surgeries (turbinate reduction, septoplasty), palatal surgery including uvulopalatoplasty, supraglottoplasty, rapid maxillary expansion, maxillary/mandibular advancement, and tracheostomy [33]. Hypoglossal nerve stimulator is a treatment option that is currently only approved for children with Down Syndrome [34–36].

Non-surgical options include positive airway pressure (PAP) therapy, medications such as montelukast, intranasal steroids, and atomoxetine/oxybutynin, weight loss, positional therapy, and watchful waiting. Continuous positive airway pressure (CPAP) offers a fixed continuous pressure that is higher than the glottic closing pressure and thereby maintains airway patency during sleep. Bilevel PAP (BPAP) consists of a higher inspiratory pressure and a lower expiratory pressure, which together helps support ventilation. It is the preferred modality for children at risk for hypoventilation, including those with neuromuscular disease and obesity. Poor adherence is the biggest limitation to PAP therapy in children. In a prospective study on PAP compliance in children, a dropout rate of 35% was recorded, alongside a low average nightly use of 5.3 h per night [31]. Skin irritation, aerophagia, and midface hypoplasia are other notable risks that are associated with PAP therapy. In a study involving children between the ages of 2 and 10 over a 16-week period, montelukast improved the severity of OSA and in a separate experiment, significantly reduced adenotonsillar size following 3 months of treatment [37] [38]. Montelukast monotherapy has been attributed to decreases in hypoxemia and respiratory disturbance, but in conjunction with intranasal steroids, leukotriene inhibitors such as montelukast can produce greater statistically significant decreases in the apnea/hypopnea index (AHI) [39] [40]. Use of only intranasal corticosteroids like budesonide for 6 weeks demonstrated improved AHIs and quality of life, a possible result of adenotonsillar proliferation and inflammation inhibition mechanisms [41]. Leukotriene inhibitors and intranasal corticosteroids for OSA treatment yield comparable AHI reduction, but psychological side effects of montelukast may pose intranasal corticosteroids as a better candidate [42]. It is important to note that these results are not reflected in obese children or children over 7–8 years old [37]. In addition, combination of atomoxetine and oxybutynin has exhibited positive effects on the severity of OSA by improving airway hypotonia, providing a potentially effective treatment for OSA in children with Down syndrome [43, 44]. The use of supplemental oxygen either via a low-flow nasal cannula or humidified high flow nasal cannula (HFNC) have been shown to be beneficial in the management of OSA, particularly in infants and young children [45–48]. Positional therapy has been shown to be beneficial among children with obesity and among those with Down syndrome [49–52]. Weight loss can play a crucial role in managing OSA in obese children [5, 53, 54]. Bariatric surgery can serve as a significant intervention when previous weight management strategies, such as dietary changes, remain ineffective [55]. By substantially reducing body weight, bariatric surgery addresses one of the primary risk factors for OSA, potentially leading to a moderation in the severity of the condition [56]. For patients with moderate or severe OSA and obesity, the use of medications like tirzepatide has shown promise in decreasing the AHI and improving sleep [57]. Data on the use of these weight loss medications in children is still lacking.

Prognosis

The prognosis for children diagnosed with OSA is, in general, favorable but it depends on the severity of the condition, presence of comorbidities, and adherence to treatment. Consequences of untreated sleep apnea include impaired mood, behavior, quality of life, poor academic performance, neurocognitive dysfunction, increased susceptibility to certain infections, elevated inflammatory markers, and cardiometabolic derangements including systemic and pulmonary hypertension, cardiac remodeling, insulin resistance, and hyperlipidemia [58–66]. Complications of adenotonsillectomy, which is the first-line treatment of pediatric OSA, include postoperative pain, bleeding, damage to surrounding tissues, and rarely velopharyngeal insufficiency and nasopharyngeal stenosis [67, 68]. Adverse effects of PAP therapy include skin irritation, aerophagia, and midface hypoplasia [69]. Data from longitudinal studies suggest that children with OSA experience poorer grades in school, lower educational levels, lower rates of employment as adults, greater healthcare costs, and higher morbidity and mortality [70, 71].

Conclusion

In conclusion, OSA in children presents a complex challenge that entails comprehensive management due to its diverse impact on health. Effective treatment and early diagnosis are crucial for mitigating these risks and improving outcomes. Although polysomnography is the gold standard for the diagnosis of OSA, it can be uncomfortable, particularly for young children or those with neurodevelopmental delays such as autism. Access to polysomnography is also a challenge for many families. Home sleep apnea testing offers a potential solution, but more data is needed in children before they can become the standard of care. With regard to treatment modalities, adenotonsillectomy continues to be the first-line treatment in children. However, with rising rates of obesity, more children are experiencing residual OSA following adenotonsillectomy. Positive airway pressure (PAP) therapy is effective for those with residual OSA, but it is not without risks. Heated high-flow nasal cannula oxygen has emerged as a potential therapeutic option for young children that are intolerant to PAP therapy. Newer medications such as atomoxetine/oxybutynin and the obesity drugs, such liraglutide, tirzepatide, and semaglutide, all hold promise in this regard. With continued advancements in understanding and treatment, the goal remains to enhance the quality of life for children affected by OSA and to prevent long-term complications.

Clinical Pearls

- Children with OSA can frequently present with irritability and hyperactivity rather than the classic daytime sleepiness seen in adults, making diagnosis challenging.
- A high index of suspicion and a thorough sleep history are crucial for early recognition of OSA in children. Children with craniofacial anomalies (Pierre-Robin Sequence), neuromuscular disease, genetic abnormalities (Down Syndrome, Prader-Willi Syndrome), and pulmonary disease (asthma, cystic fibrosis) are at increased risk for OSA and should therefore be closely monitored for the development of symptoms.
- While an in-laboratory polysomnogram (PSG) is the gold standard for diagnosing OSA, screening tools such as the Pediatric Sleep Questionnaire (PSQ-SRBD) may be a useful first step for identifying at-risk children in the primary care setting. Although home sleep apnea testing (HSAT) is widely used to diagnose adults with OSA, these are still not considered the standard of care in children.
- Obesity is an important risk factor for OSA in children, and weight management should be incorporated into the treatment algorithm for these individuals.
- Adenotonsillectomy remains the first-line treatment for children with significant adenotonsillar hypertrophy, though children with obesity or craniofacial anomalies may have residual OSA despite surgical intervention and require additional treatments such as positive airway pressure (PAP) therapy.
- Adherence to PAP therapy can be challenging in children, so clinicians should proactively address issues like mask discomfort and aerophagia to improve compliance.
- Watchful waiting, positional therapy, and medical management with intranasal corticosteroids may be considered in children with mild OSA.

References

1. Bixler EO, Vgontzas AN, Lin HM, Liao D, Calhoun S, Vela-Bueno A, et al. Sleep disordered breathing in children in a general population sample: prevalence and risk factors. Sleep. 2009;32(6):731–6. https://doi.org/10.1093/sleep/32.6.731.
2. American Academy of Sleep Medicine. International classification of sleep disorders, third edition, text revision (ICSD-3-TR). American Academy of Sleep Medicine; 2023. ISBN 978-0-9657220-9-4
3. Gozal D, Wang M, Pope DW Jr. Objective sleepiness measures in pediatric obstructive sleep apnea. Pediatrics. 2001;108(3):693–7. https://doi.org/10.1542/peds.108.3.693.
4. Perfect MM, Archbold K, Goodwin JL, Levine-Donnerstein D, Quan SF. Risk of behavioral and adaptive functioning difficulties in youth with previous and current sleep disordered breathing. Sleep. 2013;36(4):517–525B. https://doi.org/10.5665/sleep.2536.
5. Bitners A, Arens R. Evaluation and Management of Children with obstructive sleep apnea syndrome. Lung. 2020;198(2):257–70. https://doi.org/10.1007/s00408-020-00342-5.
6. Johnson C, Leavitt T, Daram SP, Johnson RF, Mitchell RB. Obstructive sleep apnea in underweight children. Otolaryngol Head Neck Surg. 2022;167(3):566–72. https://doi.org/10.1177/01945998211058722.
7. Smith DF, Amin RS. OSA and cardiovascular risk in pediatrics. Chest. 2019;156(2):402–13. https://doi.org/10.1016/j.chest.2019.02.011.
8. Redline S, Tishler PV, Schluchter M, Aylor J, Clark K, Graham G. Risk factors for sleep-disordered breathing in children. Associations with obesity, race, and respiratory problems. Am J Respir Crit Care Med. 1999;159(5 Pt 1):1527–32. https://doi.org/10.1164/ajrccm.159.5.9809079.
9. Caliendo C, Femiano R, Umano GR, Martina S, Nucci L, Perillo L, Grassia V. Effect of obesity on the respiratory parameters in children with obstructive sleep apnea syndrome. Children (Basel). 2023;10(12):1874. https://doi.org/10.3390/children10121874.
10. Gouthro K, Slowik JM. Pediatric obstructive sleep apnea. In: StatPearls [Internet]. Treasure Island (FL), StatPearls Publishing; 2024. https://www.ncbi.nlm.nih.gov/books/NBK557610/.
11. Inselma LS, Milanese A, Deurloo A. Effect of obesity on pulmonary function in children. Pediatr Pulmonol. 1993;16(2):130–7. https://doi.org/10.1002/ppul.1950160209.
12. Li AM, Chan D, Wong E, Yin J, Nelson EA, Fok TF. The effects of obesity on pulmonary function. Arch Dis Child. 2003;88(4):361–3. https://doi.org/10.1136/adc.88.4.361.
13. Arens R, Muzumdar H. Childhood obesity and obstructive sleep apnea syndrome. J Appl Physiol (1985). 2010;108(2):436–44. https://doi.org/10.1152/japplphysiol.00689.2009.
14. Gleadhill IC, Schwartz AR, Schubert N, Wise RA, Permutt S, Smith PL. Upper airway collapsibility in snorers and in patients with obstructive hypopnea and apnea. Am Rev Respir Dis. 1991;143(6):1300–3. https://doi.org/10.1164/ajrccm/143.6.1300.
15. Kang K, Chou C, Weng W, Lee P, Hsu W. Associations between adenotonsillar hypertrophy, age, and obesity in children with obstructive sleep apnea. PLoS One. 2013;8(10):e78666. https://doi.org/10.1371/journal.pone.0078666.
16. Marcus CL. Pathophysiology of childhood obstructive sleep apnea: current concepts. Respir Physiol. 2000;119(2–3):143–54. https://doi.org/10.1016/s0034-5687(99)00109-7.
17. Lumeng JC, Chervin RD. Epidemiology of pediatric obstructive sleep apnea. Proc Am Thorac Soc. 2008;5(2):242–52. https://doi.org/10.1513/pats.200708-135MG.
18. Schwengel DA, Dalesio NM, Stierer TL. Pediatric obstructive sleep apnea. Anesthesiol Clin. 2014;32(1):237–61. https://doi.org/10.1016/j.anclin.2013.10.012.
19. Savini S, Ciorba A, Bianchini C, Stomeo F, Corazzi V, Vicini C, Pelucchi S. Assessment of obstructive sleep apnoea (OSA) in children: an update. Acta Otorhinolaryngol Ital. 2019;39(5):289–97. https://doi.org/10.14639/0392-100X-N0262.
20. ElMallah M, Bailey E, Trivedi M, Kremer T, Rhein LM. Pediatric obstructive sleep apnea in high-risk populations: clinical implications. Pediatr Ann. 2017;46(9):e336–9. https://doi.org/10.3928/19382359-20170815-01.

21. Subramanyam R, Tapia IE, Zhang B, Mensinger JL, Garcia-Marcinkiewicz A, Jablonka DH, et al. Secondhand smoke exposure and risk of obstructive sleep apnea in children. Int J Pediatr Otorhinolaryngol. 2020;130:109807. https://doi.org/10.1016/j.ijporl.2019.109807.
22. Scalzitti N, Hansen S, Maturo S, Lospinoso J, O'Connor P. Comparison of home sleep apnea testing versus laboratory polysomnography for the diagnosis of obstructive sleep apnea in children. Int J Pediatr Otorhinolaryngol. 2017;100:44–51. https://doi.org/10.1016/j.ijporl.2017.06.013.
23. Tan H, Kheirandish-Gozal L, Gozal D. Pediatric home sleep apnea testing: slowly getting there. Chest. 2015;148(6):1382–95. https://doi.org/10.1378/chest.15-1365.
24. Pavone M, Ullmann N, Verrillo E, Vincentiis G, Sitzia E, Cutrera R. At-home pulse oximetry in children undergoing adenotonsillectomy for obstructive sleep apnea. Eur J Pediatr. 2017;176(4):493–9. https://doi.org/10.1007/s00431-017-2868-1.
25. Chervin RD, Hedger K, Dillon JE, Pituch KJ. Pediatric sleep questionnaire (PSQ): validity and reliability of scales for sleep-disordered breathing, snoring, sleepiness, and behavioral problems. Sleep Med. 2000;1(1):21–32. https://doi.org/10.1016/s1389-9457(99)00009-x.
26. Franco RA Jr, Rosenfeld RM, Rao M. First place–resident clinical science award 1999. Quality of life for children with obstructive sleep apnea. Otolaryngol Head Neck Surg. 2000;123(1 Pt 1):9–16. https://doi.org/10.1067/mhn.2000.105254.
27. Marcus CL, Brooks LJ, Draper KA, Gozal D, Halbower AC, Jones J, et al. Diagnosis and management of childhood obstructive sleep apnea syndrome. Pediatrics. 2012;130(3):576–84. https://doi.org/10.1542/peds.2012-1671.
28. Mitchell RB, Archer SM, Ishman SL, Rosenfeld RM, Coles S, Finestone SA, et al. Clinical practice guideline: tonsillectomy in children (update)-executive summary. Otolaryngol Head Neck Surg. 2019;160(2):187–205. https://doi.org/10.1177/0194599818807917.
29. Marcus CL, Moore RH, Rosen CL, Giordani B, Garetz SL, Taylor HG, et al. Childhood Adenotonsillectomy Trial (CHAT). A randomized trial of adenotonsillectomy for childhood sleep apnea. N Engl J Med. 2013;368(25):2366–76. https://doi.org/10.1056/NEJMoa1215881.
30. Wang R, Bakker JP, Chervin RD, Garetz SL, Hassan F, Ishman SL, et al. Pediatric Adenotonsillectomy Trial for Snoring (PATS): protocol for a randomised controlled trial to evaluate the effect of adenotonsillectomy in treating mild obstructive sleep-disordered breathing. BMJ Open. 2020;10(3):e033889. https://doi.org/10.1136/bmjopen-2019-033889.
31. Marcus CL, Rosen G, Ward SL, Halbower AC, Sterni L, Lutz J, et al. Adherence to and effectiveness of positive airway pressure therapy in children with obstructive sleep apnea. Pediatrics. 2006;117(3):e442–51. https://doi.org/10.1542/peds.2005-1634.
32. Cielo CM, Gungor A. Treatment options for pediatric obstructive sleep apnea. Curr Probl Pediatr Adolesc Health Care. 2016;46(1):27–33. https://doi.org/10.1016/j.cppeds.2015.10.006.
33. Ersu R, Chen ML, Ehsan Z, Ishman SL, Redline S, Narang I. Persistent obstructive sleep apnoea in children: treatment options and management considerations. Lancet Respir Med. 2023;11(3):283–96. https://doi.org/10.1016/S2213-2600(22)00262-4.
34. Caloway CL, Diercks GR, Keamy D, Guzman V, Soose R, Raol N, Shott SR, Ishman SL, Hartnick CJ. Update on hypoglossal nerve stimulation in children with down syndrome and obstructive sleep apnea. Laryngoscope. 2020;130(4):E263–7. https://doi.org/10.1002/lary.28138.
35. Stenerson ME, Yu PK, Kinane TB, Skotko BG, Hartnick CJ. Long-term stability of hypoglossal nerve stimulation for the treatment of obstructive sleep apnea in children with down syndrome. Int J Pediatr Otorhinolaryngol. 2021;149:110868. https://doi.org/10.1016/j.ijporl.2021.110868.
36. Diercks GR, Wentland C, Keamy D, Kinane TB, Skotko B, Guzman V, Grealish E, Dobrowski J, Soose R, Hartnick CJ. Hypoglossal nerve stimulation in adolescents with down syndrome and obstructive sleep apnea. JAMA Otolaryngol Head Neck Surg. 2018;144(1):37–42. https://doi.org/10.1001/jamaoto.2017.1871.

37. Kheirandish-Gozal L, Bandla HP, Gozal D. Montelukast for children with obstructive sleep apnea: results of a double-blind, randomized, placebo-controlled trial. Ann Am Thorac Soc. 2016;13(10):1736–41. https://doi.org/10.1513/AnnalsATS.201606-432OC.
38. Goldbart AD, Greenberg-Dotan S, Tal A. Montelukast for children with obstructive sleep apnea: a double-blind, placebo-controlled study. Pediatrics. 2012;130(3):e575–80. https://doi.org/10.1542/peds.2012-0310.
39. Montelukast for Sleep Apnea. A review of the clinical effectiveness, cost effectiveness, and guidelines [Internet]. Ottawa: Canadian Agency for Drugs and Technologies in Health; 2014. https://www.ncbi.nlm.nih.gov/books/NBK195647/
40. Liming BJ, Ryan M, Mack D, Ahmad I, Camacho M. Montelukast and nasal corticosteroids to treat pediatric obstructive sleep apnea: a systematic review and meta-analysis. Otolaryngol Head Neck Surg. 2019;160(4):594–602. https://doi.org/10.1177/0194599818815683.
41. Gudnadottir G, Ellegård E, Hellgren J. Intranasal budesonide and quality of life in pediatric sleep-disordered breathing: a randomized controlled trial. Otolaryngol Head Neck Surg. 2018;158(4):752–9. https://doi.org/10.1177/0194599817742597.
42. Kovesi T. Neuropsychiatric side effects of montelukast. J Pediatr. 2019;212:248. https://doi.org/10.1016/j.jpeds.2019.05.019.
43. Combs D, Edgin J, Hsu CH, Bottrill K, Van Vorce H, Gerken B, et al. The combination of atomoxetine and oxybutynin for the treatment of obstructive sleep apnea in children with down syndrome. J Clin Sleep Med. 2023;19(12):2065–73. https://doi.org/10.5664/jcsm.10764.
44. Aishah A, Loffler KA, Toson B, Mukherjee S, Adams RJ, Altree TJ, et al. One month dosing of Atomoxetine plus oxybutynin in obstructive sleep apnea: a randomized, Placebo-controlled Trial. Ann Am Thorac Soc. 2023;20(4):584–95. https://doi.org/10.1513/AnnalsATS.202206-492OC.
45. Brockbank J, Astudillo CL, Che D, Tanphaichitr A, Huang G, Tomko J, et al. Supplemental oxygen for treatment of infants with obstructive sleep apnea. J Clin Sleep Med. 2019;15(8):1115–23. https://doi.org/10.5664/jcsm.7802.
46. Ignatiuk D, Schaer B, McGinley B. High flow nasal cannula treatment for obstructive sleep apnea in infants and young children. Pediatr Pulmonol. 2020;55(10):2791–8.
47. Kwok K, Lau M, Leung S, Ng D. Use of heated humidified high flow nasal cannula for obstructive sleep apnea in infants. Sleep Med. 2020;74:332–7.
48. Hawkins S, Huston S, Campbell K, Halbower A. High-flow, heated, humidified air via nasal cannula treats CPAP-intolerant children with obstructive sleep apnea. J Clin Sleep Med. 2017;13(8):981–9. https://doi.org/10.5664/jcsm.6700.
49. Xiao L, Baker A, Voutsas G, Massicotte C, Wolter NE, Propst EJ, Narang I. Positional device therapy for the treatment of positional obstructive sleep apnea in children: a pilot study. Sleep Med. 2021;85:313–6. https://doi.org/10.1016/j.sleep.2021.07.036.
50. Selvadurai S, Voutsas G, Massicotte C, Kassner A, Katz SL, Propst EJ, et al. Positional obstructive sleep apnea in an obese pediatric population. J Clin Sleep Med. 2020;16(8):1295–301. https://doi.org/10.5664/jcsm.8496.
51. Lackey TG, Tholen K, Pickett K, Friedman N. Residual OSA in down syndrome: does body position matter? J Clin Sleep Med. 2023;19(1):171–7. https://doi.org/10.5664/jcsm.10288.
52. Ergenekon AP, Gokdemir Y, Ersu R. Medical treatment of obstructive sleep apnea in children. J Clin Med. 2023;12(15):5022. https://doi.org/10.3390/jcm12155022.
53. Andersen IG, Holm JC, Homøe P. Impact of weight-loss management on children and adolescents with obesity and obstructive sleep apnea. Int J Pediatr Otorhinolaryngol. 2019;123:57–62. https://doi.org/10.1016/j.ijporl.2019.04.031.
54. Van Hoorenbeeck K, Verhulst SL. Metabolic complications and obstructive sleep apnea in obese children: time to wake up! Am J Respir Crit Care Med. 2014;189(1):13–5. https://doi.org/10.1164/rccm.201311-2079ED.
55. McNicholas WT, Pevernagie D. Obstructive sleep apnea: transition from pathophysiology to an integrative disease model. J Sleep Res. 2022;31(4):e13616. https://doi.org/10.1111/jsr.13616.

56. Wyszomirski K, Walędziak M, Różańska-Walędziak A. Obesity, bariatric surgery and obstructive sleep apnea-a narrative literature review. Medicina (Kaunas). 2023;59(7):1266. https://doi.org/10.3390/medicina59071266.
57. Malhotra A, Grunstein RR, Fietze I, Weaver TE, Redline S, Azarbarzin A, et al. Tirzepatide for the treatment of obstructive sleep apnea and obesity. N Engl J Med. 2024;391:1193. https://doi.org/10.1056/NEJMoa2404881.
58. Owens J, Opipari L, Nobile C, Spirito A. Sleep and daytime behavior in children with obstructive sleep apnea and behavioral sleep disorders. Pediatrics. 1998;102(5):1178–84. https://doi.org/10.1542/peds.102.5.1178.
59. Lewin DS, Rosen RC, England SJ, Dahl RE. Preliminary evidence of behavioral and cognitive sequelae of obstructive sleep apnea in children. Sleep Med. 2002;3(1):5–13. https://doi.org/10.1016/s1389-9457(01)00070-3.
60. Mitchell RB, Kelly J. Behavior, neurocognition and quality-of-life in children with sleep-disordered breathing. Int J Pediatr Otorhinolaryngol. 2006;70(3):395–406. https://doi.org/10.1016/j.ijporl.2005.10.020.
61. Goyal A, Pakhare AP, Bhatt GC, Choudhary B, Patil R. Association of pediatric obstructive sleep apnea with poor academic performance: a school-based study from India. Lung India. 2018;35(2):132–6. https://doi.org/10.4103/lungindia.lungindia_218_17.
62. Menzies B, Teng A, Burns M, Lah S. Neurocognitive outcomes of children with sleep disordered breathing: a systematic review with meta-analysis. Sleep Med Rev. 2022;63:101629. https://doi.org/10.1016/j.smrv.2022.101629.
63. Kheirandish-Gozal L, Gozal D. Obstructive sleep apnea and inflammation: proof of concept based on two illustrative cytokines. Int J Mol Sci. 2019;20(3):459. https://doi.org/10.3390/ijms20030459.
64. Teo DT, Mitchell RB. Systematic review of effects of adenotonsillectomy on cardiovascular parameters in children with obstructive sleep apnea. Otolaryngol Head Neck Surg. 2013;148(1):21–8. https://doi.org/10.1177/0194599812463193.
65. Shamsuzzaman A, Szczesniak RD, Fenchel MC, Amin RS. Glucose, insulin, and insulin resistance in normal-weight, overweight and obese children with obstructive sleep apnea. Obes Res Clin Pract. 2014;8(6):e584–91. https://doi.org/10.1016/j.orcp.2013.11.006.
66. Siriwat R, Wang L, Shah V, Mehra R, Ibrahim S. Obstructive sleep apnea and insulin resistance in children with obesity. J Clin Sleep Med. 2020;16(7):1081–90. https://doi.org/10.5664/jcsm.8414.
67. Uwiera TC. Considerations in surgical management of pediatric obstructive sleep apnea: tonsillectomy and beyond. Children (Basel). 2021;8(11):944. https://doi.org/10.3390/children8110944.
68. De Luca CG, Pachêco-Pereira C, Aydinoz S, Bhattacharjee R, Tan HL, Kheirandish-Gozal L, Flores-Mir C, et al. Adenotonsillectomy complications: a meta-analysis. Pediatrics. 2015;136(4):702–18. https://doi.org/10.1542/peds.2015-1283.
69. Hady KK, Okorie CUA. Positive airway pressure therapy for pediatric obstructive sleep apnea. Children (Basel). 2021;8(11):979. https://doi.org/10.3390/children8110979.
70. Jennum P, Rejkjær-Knudsen M, Ibsen R, Kiær EK, von Buchwald C, Kjellberg J. Long-term health and socioeconomic outcome of obstructive sleep apnea in children and adolescents. Sleep Med. 2020 Nov;75:441–7. https://doi.org/10.1016/j.sleep.2020.08.017.
71. Blechner M, Williamson AA. Consequences of obstructive sleep apnea in children. Curr Probl Pediatr Adolesc Health Care. 2016;46(1):19–26. https://doi.org/10.1016/j.cppeds.2015.10.007.

Other Sleep Apneas

Understanding Central, Complex, and Mixed Sleep Apnea in Pediatric Populations

Janey Dudley, Daniel Rongo, and Brittney Pryor Craig

Central sleep apnea (CSA) is a disorder in which the central nervous system fails to initiate respiratory effort during sleep [1]. This condition is different from obstructive sleep apnea (OSA), where respiratory effort continues despite airway obstruction [2]. The diagnosis of CSA is typically confirmed through polysomnography (PSG), which shows episodes of apnea and oxygen desaturation without respiratory effort. The central apnea index (CAI) is used to measure the severity of CSA. Though no one official severity scale for sleep apnea in children exists, a widely accepted consensus is that a CAI of 5 or more events per hour suggests the presence of clinically significant central sleep apnea [3]. Symptoms of CSA may manifest differently for children than they do for adults. Symptoms can include excessive daytime sleepiness, behavioral issues such as hyperactivity, irritability, or mood swings, and decreased cognitive function. CSA affects less than 5% of healthy children [1]. The causes of CSA are often multifactorial and commonly linked to conditions such as congestive heart failure, stroke, or other neurological disorders that affect the brain's ability to regulate breathing. For example, in infants and children, CSA may be associated with congenital conditions affecting the central nervous system, such as Chiari malformations or congenital central hypoventilation syndrome (CCHS). There are several types of Chiari malformations, but Chiari type 1 malformation is the most common [4]. A Chiari type 1 malformation is a structural brain defect that occurs when a portion of the cerebellum, which is situated in the back part of the skull, is displaced down through the foramen magnum into the spinal canal. This can put pressure on the brainstem and spinal cord which, when symptomatic,

---

J. Dudley · D. Rongo
Neurology Department, Sleep Medicine Fellowship, UT Southwestern, Dallas, TX, USA
e-mail: janey.dudley@utsouthwestern.edu; daniel.rongo@utsouthwestern.edu

B. Pryor Craig (✉)
Departments of Neurology and Sleep Medicine, Cook Children's Medical Center, Prosper, TX, USA
e-mail: Brittney.PryorCraig@cookchildrens.org

manifests in multiple symptoms including headaches, voice changes, nystagmus, difficulty swallowing, disturbance in extremity sensation, gait changes, and obstructive, and/or central sleep apnea [4]. Congenital central hypoventilation syndrome is a rare genetic condition caused by a mutation in the PHOX2B gene which results in malfunction of the nerves that control involuntary body functions, particularly breathing [5]. Therefore, individuals with CCHS exhibit central apneas during sleep. Central sleep apnea is less prevalent than OSA, but can be high risk and difficult to isolate in certain populations. For instance, periodic breathing is a benign, common occurrence in preterm and term infants where immature respiratory control can lead to episodes of central apneas. This can be challenging to delineate from pathologic CSA in newborns without using PSG [1]. Clinicians should have higher clinical suspicion for the possibility of pathologic CSA in special populations at increased risk including children with neurological impairments, those on central nervous system depressants, and patients with severe heart or renal failure. Treatment options for CSA vary based on the underlying etiology. In cases of neonatal periodic breathing, watchful waiting may be appropriate since many infants outgrow this condition. For altitude-induced central apnea (which is due to an overactive physiologic respiratory response that can be seen even in healthy individuals ascending to high altitudes), supplemental oxygen and medications like acetazolamide (Diamox) may be effective. In more severe cases of CCHS, tracheostomy and mechanical ventilation are often required to ensure adequate ventilation during sleep. Overall, behavioral interventions and parental education play a crucial role in managing CSA in pediatric patients, particularly in those with underlying neurological conditions (Fig. 10.1).

Complex sleep apnea (CompSA) is a specific form of CSA that emerges following the treatment of obstructive sleep apnea, such as after tonsillectomy and adenoidectomy (T&A) or the use of continuous positive airway pressure (CPAP) therapy. This phenomenon occurs when central apneas develop after the reduction or resolution of the obstructive events. Clinicians should suspect CompSA in patients who initially show improvement in symptoms but later experience a recurrence of daytime sleepiness or other symptoms after surgical intervention or while

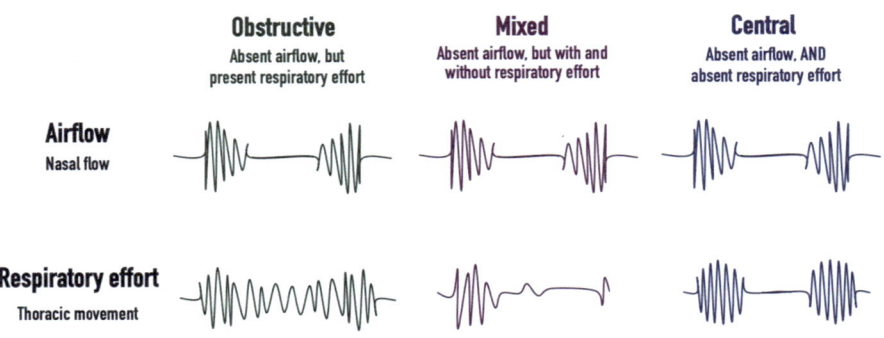

Fig. 10.1 Apneas may be obstructive, mixed, or central

on CPAP therapy. Recognizing CompSA can be particularly challenging since it sometimes mimics the original symptoms of OSA, complicating the diagnostic process. The workup for CompSA involves repeating the PSG, which would demonstrate an improvement in obstructive events with a worsening of central apneas. PSG is critical in diagnosis, as it helps differentiate between persistent OSA and the emergence of CompSA. Treatment strategies may include continued use of CPAP, which can sometimes lead to spontaneous resolution of central apneas over time. If central apneas persist, bilevel positive airway pressure (BiPAP) or bilevel positive airway pressure with spontaneous-timed mode (BiPAP ST) may be employed to provide sufficient ventilation. In some cases, a multidisciplinary approach involving sleep specialists, pulmonologists, and neurologists may be necessary to optimize management and address any underlying conditions contributing to the emergence of CompSA [6, 7].

A mixed apnea (MA) is a distinct respiratory event during sleep that has characteristics of both obstructive and central apneas with respiratory effort being present in one part and absent in another part. This is not to be confused with mixed sleep apnea (MSA) which refers to the specific type of sleep-disordered breathing in which both obstructive and central sleep apneas appear at clinically significant severities; mixed apneas may also be present. The prevalence of MSA in children is not well-defined, but it is recognized as a separate clinical entity that can complicate the management of sleep-disordered breathing. Mixed sleep apnea may occur in children with a history of persistent OSA who subsequently develop central apneas, often as a result of treatment interventions or underlying medical conditions. The treatment of MSA often focuses on addressing the obstructive component, since effective management of OSA may lead to improvements in the central apnea component. An example of this is when interventions such as T&A for OSA may also alleviate some of the central apneas that were seen on initial PSG, highlighting the interconnected nature of these disorders. In cases where MSA is diagnosed, a comprehensive treatment plan may include surgical interventions, continuous positive airway pressure (CPAP) therapy, lifestyle modifications, and regular follow-up assessments to monitor the child's progress and to adjust treatment, as necessary [8].

In conclusion, understanding the nuances of central, complex, and mixed sleep apnea is crucial for effective diagnosis and management in pediatric populations. Each type presents unique challenges and requires tailored treatment strategies to optimize patient outcomes. Ongoing research into the pathophysiology, epidemiology, and treatment options for these conditions will further enhance our understanding and improve the care provided to affected children.

Clinical Vignette

Nick is a 7-year-old male who was referred to the sleep clinic with complaints of persistent daytime sleepiness and behavioral problems despite a recent tonsillectomy and adenoidectomy (T&A) performed to treat moderate obstructive sleep apnea (OSA). His initial pre-operative polysomnogram (PSG) showed an obstructive apnea-hypopnea index (oAHI) of 6.0/h and central apnea index (CAI) of 0.0/h with an oxygen saturation nadir of 88%. His parents reported that his snoring and noisy breathing improved post-surgery, but he is still groggy upon awakening, and his teacher said that Nick has been dozing off in the mornings and having difficulty staying seated and following directions in the afternoons.

Pertinent findings on exam: Nick appeared awake but was frequently inattentive. The nares were patent, and there was no nasal congestion. The oropharynx was clear and tonsils were absent. Heart rate was normal. Heart rhythm was regular. There were no murmurs. Lungs were clear to auscultation bilaterally.

A repeat PSG was conducted to further investigate for possible residual OSA. The PSG revealed an oAHI of 0.5/h and a CAI of 10.2/h with an oxygen saturation nadir of 92%. The sleep study confirmed that the T&A had successfully treated the OSA, but Nick had subsequently developed a significant number of treatment-emergent central apneas consistent with a diagnosis of complex sleep apnea (CompSA).

Given the new diagnosis of CompSA, which was believed to be the cause of his daytime symptoms, the management plan was adjusted. A trial of continuous positive airway pressure (CPAP) was initiated; however, the central apneas persisted, so he was transitioned to bilevel positive airway pressure with spontaneous-timed mode (BiPAP ST) to provide optimal ventilation. Nick was also referred to a multidisciplinary team that included a sleep specialist, pulmonologist, and neurologist to evaluate for and address any underlying conditions that could be contributing to his presentation. No cardiopulmonary etiology was found. A small Chiari 1 malformation was discovered on brain MRI. In consultation with a neurologist and neurosurgeon, family elected against decompression surgery and to keep close surveillance with continued conservative management of his CompSA with BiPAP ST as prescribed by the sleep specialist.

During follow-up visits, Nick and parents reported marked improvement in his daytime sleepiness and behavioral issues as his central apneas were well-managed with BiPAP ST, which he continued for 1.5 years after which his CompSA resolved as demonstrated on repeat PSG. This case highlights the importance of recognizing the potential for complex sleep apnea to develop after the successful treatment of obstructive sleep apnea and the need for a tailored, multidisciplinary approach in managing such cases.

Clinical Pearls

- Central sleep apnea (CSA) is a disorder in which the central nervous system fails to initiate respiratory effort during sleep.
- Symptoms may manifest differently for children than they do for adults, including excessive daytime sleepiness, behavioral issues such as hyperactivity, irritability, or mood swings, and decreased cognitive function.
- The causes of CSA are often multifactorial and commonly linked to conditions such as congestive heart failure, stroke, or other neurological disorders that affect the brain's ability to regulate breathing.

- Mixed sleep apnea (MSA) is when central sleep apnea presents concurrently with obstructive sleep apnea.
- Complex sleep apnea (CompSA) is a specific form of CSA that emerges following the treatment of obstructive sleep apnea, such as after tonsillectomy and adenoidectomy (T&A) or the use of continuous positive airway pressure (CPAP) therapy.
- Treatment options of CSA, MSA, and CompSA are multifactorial and can vary greatly between children with similar pathology, often requiring a multidisciplinary approach.

References

1. Ghirardo S, Amaddeo A, Griffon L, Khirani S, Fauroux B. Central apnea and periodic breathing in children with underlying conditions. J Sleep Res. 2021;30(6):e13388. https://doi.org/10.1111/jsr.13388.
2. Sans Capdevila O, Kheirandish-Gozal L, Dayyat E, Gozal D. Pediatric obstructive sleep apnea. Proc Am Thorac Soc. 2008;5:274. https://doi.org/10.1513/pats.200708-138MG.
3. Mclaren AT, Bin-hasan S, Narang I. Diagnosis, management and pathophysiology of central sleep apnea in children. Paediatr Respir Rev. 2019;30:49–57. https://doi.org/10.1016/j.prrv.2018.07.005.
4. Ellenbogen RG, et al. Chiari malformations and syringohydromyelia. In: Principles of neurological surgery. 4th ed. Elsevier; 2017. https://www.clinicalkey.com. Accessed 27 Nov 2024.
5. Zaidi S, Gandhi J, Vatsia S, Smith NL, Khan SA. Congenital central hypoventilation syndrome: an overview of etiopathogenesis, associated pathologies, clinical presentation, and management. Auton Neurosci. 2018;210:1–9. https://doi.org/10.1016/j.autneu.2017.11.003. Epub 2017 Nov 13. PMID: 29249648.
6. Gay P. Complex sleep apnea: it really is a disease. J Clin Sleep Med. 2008;4(5):403. https://doi.org/10.5664/jcsm.27272.
7. Wollin D, Castro Codesal M, DeHaan K, MacLean J. Characterizing treatment emergent central sleep apnea in children. Sleep. 2017;40(1):a320. https://doi.org/10.1093/sleepj/zsx050.860.
8. Antunes J, Carvalho J, Marinho C, et al. Central and mixed apneas in children with obstructive sleep apnea: effect of adenotonsillectomy. Eur Arch Otorhinolaryngol. 2024;281:3125. https://doi.org/10.1007/s00405-023-08442-7.

Parasomnias in Children

11

David Rongo, Daniel Rongo, and Shan Luong

Introduction

Parasomnias are a group of sleep-related disorders of arousal that typically occur as undesirable physical events or experience while falling asleep (hypnogogic), during sleep, or before awakening (hypnopompic). Although parasomnias are found across all age groups, children are more likely to manifest these issues, particularly non-REM parasomnias.

Types of Parasomnias (Fig 11.1)

Parasomnias are broadly divided into three categories: REM, Non-REM, and other parasomnias (typically related to sleep-wake transitions).

Non-REM Related Parasomnias

Non-REM (NREM) parasomnias originate from incomplete arousals from deep sleep (slow wave sleep) that typically occur during the first half of the night where NREM is more predominant. During slow-wave sleep, higher brain functions are suppressed, which often leads to patients not remembering these occurrences.

D. Rongo
Department of Pediatrics, Drexel University/St. Christopher's Hospital for Children, Philadelphia, PA, USA
e-mail: David.Rongo@TowerHealth.org

D. Rongo (✉)
Department of Neurology, UT Southwestern, Dallas, TX, USA
e-mail: Daniel.Rongo@UTSouthwestern.edu

S. Luong
Department of Internal Medicine – Pulmonary Division, UT Southwestern, Dallas, TX, USA
e-mail: shan.luong@utsouthwestern.edu

© The Author(s), under exclusive license to Springer Nature Switzerland AG 2025
A. Wani, I. S. Khawaja (eds.), *Sleep Disorders in Children*,
https://doi.org/10.1007/978-3-031-92166-7_11

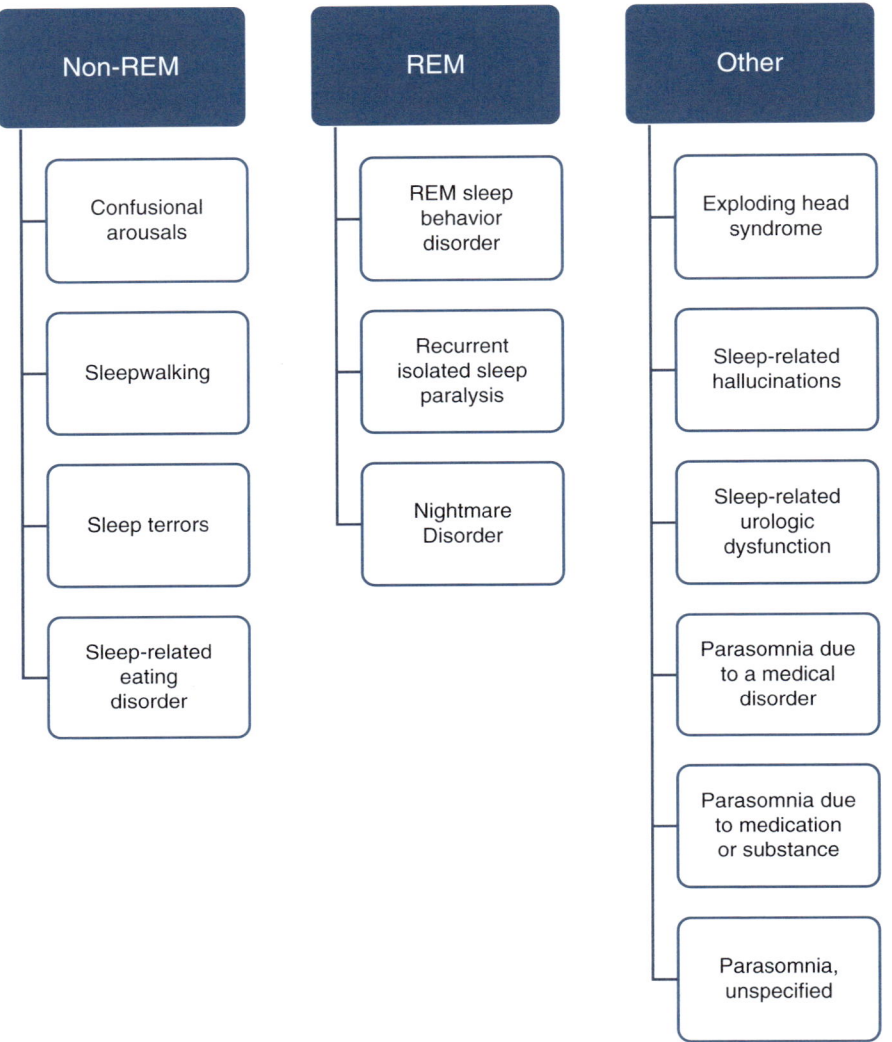

Fig. 11.1 Parasomnias classifications according to the "International Classification of Sleep Disorders"—Third Edition

Patients are unable to respond to parents or siblings who try to interact with them during these episodes. They often appear confused and disoriented for several minutes or longer following the episode. NREM parasomnias include confusional arousals, sleepwalking, sleep terrors, and sleep-related eating disorders. Significant injury can occur if parents or caregivers do not take appropriate measures to ensure the patient's safety during sleep.

Confusional arousals can include moaning, crying, or sitting up in bed with a confused appearance. However, they do not typically exhibit the features seen in sleepwalking, such as moving out of bed, or sleep terrors, which involve intense fear or panic. The prevalence of confusional arousal is approximately 17.3% among children 3–13 years of age [1].

Sleepwalking is a disorder characterized by ambulation and complex locomotor behaviors that meet the criteria for NREM parasomnias. Patients often do not remember the event. In young children, sleepwalking may involve walking or crawling around the room with a glassy-eyed expression, often completing aimless or semi-purposeful movements. Accidental injury during sleepwalking can pose a considerable risk if preventive measures are not taken (e.g., falling downstairs, crashing through glass doors, etc.) Childhood sleepwalking typically ceases spontaneously by adolescence.

Sleep terrors, also known as night terrors, are characterized by sudden arousals from slow-wave sleep, typically in the first third of the night. The child may sit up abruptly, scream, and appear terrified. There is autonomic arousal, including tachycardia, tachypnea, and diaphoresis during the episode. Unlike nightmares, the child is usually difficult to console and may not recognize their parents. Episodes typically last 1–10 minutes and are followed by confusion and disorientation. The child usually has no recollection of the event the next morning. Sleep terrors are most common in children aged 3–12 years, with a peak prevalence at age 3, and often resolve spontaneously with age [2]. As many as 30% of children with sleep terrors may go on to develop sleepwalking later in childhood [3].

Sleep-related eating disorder (SRED) is characterized by recurrent episodes of involuntary eating and drinking during partial arousals from sleep. Individuals with SRED may consume unusual or inedible items and have no recollection of the event the following morning. While less common in children compared to adults, it can occur and may be associated with other sleep disorders or medications. Safety precautions, such as securing the kitchen area at night, may be necessary. SRED should be differentiated from other conditions associated with nocturnal eating or hyperphagia, such as binge eating disorder Prader-Willi syndrome, and Kleine-Levin syndrome.

REM-Related Parasomnias

REM-related parasomnias occur during rapid eye movement (REM) sleep and include REM sleep behavior disorder, recurrent isolated sleep paralysis, and nightmare disorder. These parasomnias typically occur in the latter half of the night when REM sleep is more prevalent. Unlike NREM parasomnias, individuals experiencing REM parasomnias often have a clear recall of the events. In children, nightmare disorder is the most common REM-related parasomnia, while REM sleep bhavior disorder is rare. Newborn sleep architecture is not yet fully developed, and due to underdeveloped muscle atonia, newborns may exhibit behaviors such as laughing, kicking, and crying during sleep, which are manifestations of normal sleep patterns. This is referred to as "active sleep", rather than REM sleep.

REM sleep behavior disorder typically does not affect children, although some case reports have been documented. It is more common in adults over 50 years of age and is associated with alpha-synucleinopathy neurodegenerative diseases, such as Parkinson's disease, dementia with Lewy bodies (DLB), and multiple system atrophy (MSA).

Recurrent isolated sleep paralysis typically occurs in adolescents aged 14–17 years, although cases in adults have also been reported. It manifests as a recurrent night-time awakening followed by the inability to move, lasting a few seconds to minutes, and is often accompanied by significant distress and anxiety. Sleep paralysis can be secondary to other medical conditions, such as narcolepsy, substance abuse, or mental health disorders, so it is important to rule out these conditions when evaluating a patient with these complaints. Additionally, sleep deprivation can also contribute to the occurrence of sleep paralysis.

Nightmare disorder occurs when an individual experiences repeated, distressing vivid dreams. Patients typically wake up fully alert, with a clear recollection of the dream. This disorder can cause significant impairment in the patient's quality of life and disrupt their usual sleep schedule (i.e., bedtime avoidance, daytime sleepiness, impaired functioning at school). While it is common in children, nightmare disorder should be differentiated from isolated nightmares, which are a normal part of childhood development. Sleep terrors are similar but distinguishable from nightmare disorders (see Differential Diagnosis below).

Other Parasomnias

The other parasomnias most relevant to pediatric practice include sleep-related hallucinations and sleep-related urologic dysfunction (nocturnal enuresis). Substance and medical disorder-related parasomnias should also be considered in this group when a clear association is present. Exploding head syndrome is typically seen in adults over 50 years of age and is characterized by the perception of a sudden, loud noise during the transition from wake to sleep or from sleep to wake.

Sleep-Related Hallucinations

Sleep-related hallucinations occur during the transition from wake to sleep (hypnagogic) or from sleep to wake (hypnopompic). They typically affect adolescents and young adults. These hallucinations can involve any of the senses (e.g., gustatory, olfactory, somatic, emotional, visual, tactile). They can be triggered by anxiety, sleep deprivation, or substance use (e.g., alcohol).

Sleep-Related Urologic Dysfunction

Sleep-related urologic dysfunction refers to involuntary voiding during sleep and is commonly known as bedwetting, sleep enuresis, or nocturnal enuresis. It is diagnosed in a patient who experiences these episodes at least twice a week for 3 consecutive months at the age of 5 years or older. It is considered primary if the patient has never been consistently dry for at least 6 months after the age of 5 years, and secondary if the patient was dry for at least 6 months but then resumed wetting episodes for at least 3 months, occurring at least twice a week.

Pathophysiology

Parasomnias are thought to result from incomplete transitions between wakefulness and sleep, leading to a dissociative state where elements of both states coexist. This unstable state may help explain the complex behaviors and emotions seen in parasomnias. These events often occur during arousals from sleep.

The cause of these disorders is multifactorial, with both environmental and genetic factors contributing to the clinical presentation of arousal disorders. Sleep fragmentation caused by conditions such as obstructive sleep apnea, periodic limb movement disorder, or the use of certain medications can increase the risk of parasomnias. The immaturity of sleep-wake boundaries in children may explain why parasomnias are more prevalent in younger individuals. Genetics plays a key role, as many patients with parasomnias have a family member who also experiences the condition.

Diagnosis and Evaluation

The diagnosis of parasomnia typically begins with a detailed sleep history, often with the help of a family member or loved one. Patients are commonly described as "acting out their dreams" or "doing strange things at night". Caregivers may bring a video recording to their visit, showing the patient during a specific event.

As a clinician, the next steps in evaluating a patient with suspected parasomnia include inquiring about:

- Timing and frequency of events
- Sleep schedule and duration
- Specific behaviors observed
- Triggers or patterns noticed
- Family history of sleep disorders
- Medication use
- Alcohol and substance use
- Comorbid conditions (e.g., sleep apnea, restless leg, bruxism)

Timing is a key feature in diagnosing parasomnias, as it helps differentiate between REM and non-REM parasomnias. REM parasomnias typically occur during the later phases of the night, when REM sleep is more prevalent in the early morning hours. Non-REM parasomnias are more commonly observed in the first third of the night, when deeper stages of non-REM sleep predominate.

Understanding the triggers of parasomnia is an essential part of the patient's history. These triggers can include:

1. Specific medications: Certain drugs are known to induce or exacerbate parasomnias. Examples include:
   - SSRIs (Selective Serotonin Reuptake Inhibitors)
   - Zolpidem and other sedative-hypnotics
   - Antihistamines
2. Sleep deprivation: Insufficient sleep or irregular sleep patterns can increase the likelihood of parasomnia episodes.
3. Recent fever or illness: Particularly in children, febrile illnesses can trigger parasomnia, such as night terrors or confusional arousals.
4. New or persisting emotional stressors: Psychological stress can significantly impact sleep quality and trigger various parasomnias.

Family history is also important, as patients with parasomnia often have a first-degree relative who also has the condition.

Differential Diagnosis

The differential diagnosis of parasomnias should include consideration of conditions that mimic parasomnias, such as seizures, headaches, and psychiatric conditions (e.g., panic disorder, PTSD). As mentioned previously, it is important to recognize comorbid conditions that may act as triggers for parasomnias.

Differentiating between sleep terrors (night terrors) and nightmares is important, as the mechanisms and management differ between these two conditions (Table 11.1).

Management and Treatment

Management of parasomnias typically involves a combination of non-pharmacologic behavioral interventions and, in some cases, medication. Counseling and education are key components of patient care. For certain parasomnias, specific therapies are known to be effective. Below are general principles for management:

1. Counseling/Education:
   - Explain the benign nature of most parasomnias.
   - Emphasize that most children will outgrow these.
2. Sleep Hygiene and Bedtime Routine:
   - Establishing a consistent sleep schedule.
   - Limit screen time and light exposure at night.
3. Safety Measures:
   - Ensure a safe environment to prevent injury during an episode.

Table 11.1 Differences between nightmares and sleep terrors (night terrors)

Characteristic	Nightmares	Sleep terror (Night terror)
Timing	REM sleep (second half of the night)	Non-REM sleep (first half of the night)
Recall	Usually clear recall	Little to no recall
Wakefulness	Fully awake after episode	Partial or no wakefulness during episode
Content	Frightening dream content	No specific dream content
Consolability	Can be comforted	Difficult to comfort or wake
Duration	Few minutes	10–30 min
Behavior	May seek comfort, can describe dream	Screaming, thrashing, appearing terrified
Autonomic arousal	Mild	Significant (sweating, rapid heartbeat)
Return to sleep	May have difficulty	Usually return to sleep easily
Associated factors	Stress, trauma, anxiety	Sleep deprivation, fever, certain medications
Impact on sleep quality	May affect next-day functioning	Usually doesn't affect daytime functioning
Treatment approach	Reassurance, sleep hygiene, sometimes therapy (i.e., Imagery Rehearsal Therapy)	Ensuring safety, adjusting sleep schedule (i.e., scheduled awakenings)

- Secure windows and doors to prevent escape.
- For sleepwalkers, consider door alarms or bells to notify parents if the child leaves the room.
- Remove potentially dangerous items from room.

4. Additional Considerations:
   - In patients with relevant risk factors, consider evaluating for other sleep-related conditions that may cause poor sleep quality or fragmentation, which could predispose them to parasomnias.

Specific Management

Sleep Terrors

Scheduled awakenings have shown promise in patients with sleep terrors, significantly reducing the frequency and intensity of episodes over time. Many children who undergo this intervention experience a marked decrease in night terrors, with some achieving complete cessation. This method works by disrupting the sleep cycle pattern that may contribute to the occurrence of night terrors, potentially leading to long-term improvement in sleep quality and a reduction in parasomnia events. Patients with more frequent and regular sleep terrors are often more responsive to this approach. In addition, preventing sleep deprivation and ruling out other sleep-related conditions can help reduce the occurrence of sleep terrors in the long term.

For severe, frequent, or functionally impairing sleep terrors (e.g., excessive daytime sleepiness), clonazepam—an agent that suppresses stage three of NREM sleep when taken at least 90 min before sleep—may be considered on a short-term basis. Clonazepam should be tapered gradually to avoid recurrence of sleep terrors upon discontinuation.

Most children outgrow the disorder by late adolescence and reassurance should be provided to families.

Nightmare Disorder

Most children experience nightmares at some point during their lifetime. However, when nightmares occur with increased frequency that negatively impacts a child's long-term quality of life, targeted interventions become necessary to prevent these distressing nocturnal events. In such cases, a proactive approach to nightmare prevention is not only reasonable but often essential for the child's overall well-being and development.

Image rehearsal therapy (IRT) is an effective treatment for preventing nightmares. In this approach, the clinician helps the child reimagine and rewrite the nightmare, transforming it into a less frightening version. The child practices this new, non-threatening version of the nightmare, which over time can lead to the extinction of the original nightmare. Addressing underlying causes, such as recurrent stressors in the child's life, can also be beneficial in reducing nightmares.

Nocturnal Enuresis

Careful evaluation is necessary to investigate other potential causes of involuntary nocturnal voiding. A bladder diary, documenting episodes of incontinence, can be useful for clinician assessment. Daytime incontinence may indicate lower urinary tract dysfunction, and referral for urological evaluation should be considered. Urinalysis to rule out conditions such as diabetes mellitus, diabetes insipidus, or urinary tract infection (UTI) can also be helpful.

Management options for nocturnal enuresis include both non-pharmacologic and pharmacologic treatments, which can be used alone or in combination. The use of a bed-wetting alarm is the most effective treatment. These alarms contain a moisture sensor placed in the child's undergarments, which triggers an alarm at the onset of urination. Parents should also be advised to limit the child's fluid intake before bedtime.

Desmopressin, an antidiuretic, can be helpful in children who fail non-pharmacologic treatments. Other pharmacologic options include muscarinic antagonists, such as oxybutynin, and tricyclic antidepressants, such as imipramine. Urologic referral should be considered if underlying urologic comorbidities are suspected or if the child fails to respond to standard therapy.

Follow-Up

Unhealthy sleep in children has been linked to numerous negative health outcomes. Insufficient sleep and poor sleep quality can impair neurocognitive functions in both young children and adolescents. Parasomnias are closely associated with conditions such as anxiety, depression, and ADHD. Poor sleep can lead to disruptive behaviors, negatively affecting the child's relationships with peers and family members, as well as academic performance. Close follow-up between patients and providers is essential to prevent these adverse consequences.

Pearls/Take-Home Points

- Parasomnias are thought to result from incomplete transitions between wakefulness and sleep, leading to a dissociative state where elements of both states coexist.
- Although parasomnias can affect individuals of all age groups, children are more likely to experience these disorders, particularly non-REM parasomnias.
- Parasomnias are broadly categorized into three types: REM parasomnias, Non-REM parasomnias, and other parasomnias (often related to sleep-wake transitions).
- Significant injury can occur if safety measures are not taken to protect the patient during sleep.

Patient Case

An otherwise healthy 5-year-old girl presents to her primary care clinic with her father, who is concerned about her sleep interruptions that have developed over the past 4 months. He reports that she suddenly sits up in bed, appears frightened, and cries inconsolably. She is often seen with her eyes open during these episodes but appears unresponsive and unaware of her parents' presence. The episodes typically last about 5–10 min and occur once to twice per week. He is concerned that something could be wrong and is seeking assistance. The patient was born full-term via uncomplicated delivery and has met all of her developmental milestones. She is not currently taking any prescription medications. Her father recalls having episodes of sleepwalking as a child.

Her vital signs are unremarkable, and she is in the 46th percentile for both weight and height. On examination, she appears relaxed and well-nourished. Her oropharynx is normal without crowding, and her tonsils are 1+ bilaterally. Laboratory results are unremarkable for anemia or metabolic disturbances, and thyroid function is within normal limits. She undergoes an in-lab polysomnogram, which shows an AHI of 0.5 events/hour, with no concerns for sleep apnea.

References

1. Ohayon MM, Carskadon MA. Prevalence and patterns of sleep disorders in the general population. Sleep. 2000;23(1):143–8.
2. Hublin C, Partinen M, Koskenvuo M. Sleep and sleep disorders in adults: a population-based study. Sleep. 1997;20(6):400–6.
3. Miano S, Ferri R, Bruni O. Sleepwalking in childhood: a review of the clinical characteristics, diagnosis, and treatment. Sleep Med Rev. 2007;11(4):175–85.

Sleep-Related Movement Disorders

12

Lourdes M. DelRosso

Introduction

Sleep-related movement disorders represent a significant and often under-recognized aspect of pediatric sleep medicine. These disorders encompass a variety of conditions characterized by involuntary movements that occur during sleep or at the transition between wakefulness and sleep. Such movements can range from subtle, repetitive actions to more pronounced, and disruptive behaviors, all of which can significantly impact the quality of sleep and, consequently, overall health and development. The importance of sleep in childhood cannot be overstated, as it is a critical period for physical growth, cognitive development, and emotional regulation. Disruptions in sleep due to movement disorders can lead to fragmented sleep architecture, resulting in insufficient restorative sleep. This, in turn, can manifest as daytime sleepiness, behavioral issues, impaired academic performance, and even long-term health consequences. The etiology of sleep-related movement disorders in children is multifactorial, involving genetic predispositions, neurobiological factors, and environmental influences. Understanding these underlying mechanisms is crucial for developing effective diagnostic and therapeutic strategies. Diagnosis often requires a comprehensive approach, including detailed clinical history, physical examination, and sometimes polysomnography to capture the specific movements and their impact on sleep. Management of these disorders is equally complex and necessitates a multidisciplinary approach. Treatment strategies may include pharmacological interventions and lifestyle modifications tailored to the individual needs of the child. The role of parents and caregivers is paramount in recognizing symptoms, seeking medical advice, and adhering to treatment plans. Education and support for families are essential components of effective management, helping them navigate the challenges associated with these disorders. Ultimately, the goal is to enhance the quality of life for affected children by ensuring they receive

L. M. DelRosso (✉)
University of California San Francisco, Fresno, CA, USA

© The Author(s), under exclusive license to Springer Nature Switzerland AG 2025
A. Wani, I. S. Khawaja (eds.), *Sleep Disorders in Children*,
https://doi.org/10.1007/978-3-031-92166-7_12

adequate, restorative sleep. This involves not only addressing the movement disorders themselves but also considering the broader context of the child's health and well-being. By fostering a holistic approach that integrates medical, psychological, and social support, we can better address the needs of children with sleep-related movement disorders. This section aims to provide a thorough overview of sleep-related movement disorders in children, highlighting the importance of diagnosis, effective treatment, and ongoing support to mitigate the impact of these conditions.

Sleep-Related Rhythmic Movement Disorders

Sleep-related rhythmic movement disorders (SRRMD) comprise a group of movement disorders that are characterized by repetitive, stereotyped, and rhythmic motor behaviors that occur predominantly early in childhood with an average age of onset of 9 months of age, although can also be seen in adults. SRRMD are characterized under Movement Disorder in the International Classification of Sleep Disorders, third Edition (ICSD-3) [1]. The most common categories of SRRMD are head banging and body rocking. Head banging is characterized by repetitive lifting of the head with banging the head into the pillow or mattress. Body rocking consists of a rhythmic forward and backward movement of the trunk that can occur in the sitting or quadruped position. Other less common SRRMD types include head rolling, leg banging, body rolling, or leg rolling. On rare occasions, several forms of rhythmic movements are seen in a single patient [2]. The prevalence at 9 months of age is up to 66%, decreased to 5% at age 5, with most cases resolving by age 10, nevertheless It can persist into adulthood [3, 4].

Pathophysiology

The pathophysiology of SRRMD is unknown, but there are several postulated mechanisms discussed below.

Self-Soothing
It has been suggested that rhythmic movements constitute a positive vestibular stimulation for self-soothing effect or a stimulus for motor development in the early stages of life [5]. This hypothesis is supported by the fact the rhythmic movements during sleep are common in the early years of life and self-resolve without any adverse consequences.

Sleep Instability
PSG findings have demonstrated a close relationship between SRRMD and arousals, leading to studies exploring sleep instability in patients with SRRMD. Cyclic alternating pattern (CAP) is the periodic electroencephalographic (EEG) activity occurring during NREM sleep. CAP is a marker of sleep instability and has standardized rules and criteria for its visual scoring [6]. A CAP cycle in composed of

phase A and phase B. Phase A is subdivided into A1, A2, and A3 phases. Phase A represents neural synchronization and has been associated with various movements during sleep (periodic limb movements, bruxism and sleepwalking) [7, 8]. Manni et al. showed that rhythmic movements during sleep occurred shortly after phase A regardless of the sleep stage [9]. Similarly, a case report of a 9 year old with SRRMD showed a relationship between phases A2 and A3 of the cycling alternating pattern and rhythmic movements during NREM sleep [10].

Central Pattern Generators
The basal ganglia may play a role in rhythmic movements during sleep [11]. The basal ganglia forms part of a system called "central pattern generators" (CPG). CPG have been hypothesized to play a role in motor phenomena during sleep including leg movements and parasomnias. The CPG is a neural network involved in the control of early motor function thought to be under the inhibitory control of the cortex. It has been postulated that immaturity of the inhibitory cortical system in early infancy might be associated with rhythmic movements during sleep in the first year of life [12].

Genetics
Familial case series have been reported in the literature. The exact genetics is unknown [13].

Clinical Features and Comorbidities

Usually, a benign self-limiting condition seen in the majority of cases in normally developing children. However, there are some associations with comorbidities such as ADHD [14, 15], autism spectrum disorder, mental retardation and in children with psychiatric disorders. [15] Rhythmic movements as explained above, by themselves do not constitute a disorder unless associated with sleep disruption, daytime sleepiness or injuries. When reported for the first time in adolescents, [16] a full work up including laboratory, brain computed tomography, magnetic resonance image, and resting electroencephalography is recommended.

In a cohort of adults with sleep-disordered breathing, 81.4% of the rhythmic movements were triggered by a respiratory event, but rarely SRRMD improved after CPAP [17].

Diagnosis

The diagnosis of SRRMD is mainly clinical and based on history, physical exam, and fulfillment of the diagnostic criteria. During sleep, patients who exhibit rhythmic movements are unresponsive to commands but if awakened they are alert, oriented, and deny dream recall. Patients can present with one type of rhythmic movement, a couple of co-existing different types which or resolution of one type with persistence of another type of movement as the patient grows older. The clinical presentation is usually mild and sporadic. It can consist of single episodes of rhythmic movements occurring at bedtime before sleep onset. Rarely persistent movements through the night or more severe forms exist.

Cases of atypical SRRMD, with head slapping or head punching have been reported both in adults and children [18, 19]. Due to these atypical presentations and the stereotypic nature of the movements, a differential diagnosis of epilepsy is often considered and in some cases patients with SRRMD warrant a referral to rule out seizures [4].

Differential Diagnosis

The differential diagnosis summarized includes parasomnias, other movement disorders such as periodic limb movement disorder, rapid eye movement-sleep behavior disorder, and sleep myoclonus [20]. The majority of times the diagnosis can be made with a clinical evaluation alone but when atypical features are present, video-polysomnography will aid in the diagnosis.

Clinical and Laboratory Evaluation

The features of rhythmic movement seen by home video recording and a full-night PSG examination may be useful (Fig. 12.1), particularly in differentiating SRRMD from other sleep-related diseases (bruxism, PLMS) or in identifying co-morbid sleep disorders (OSA). Full night polysomnography has showed that SRRMD can

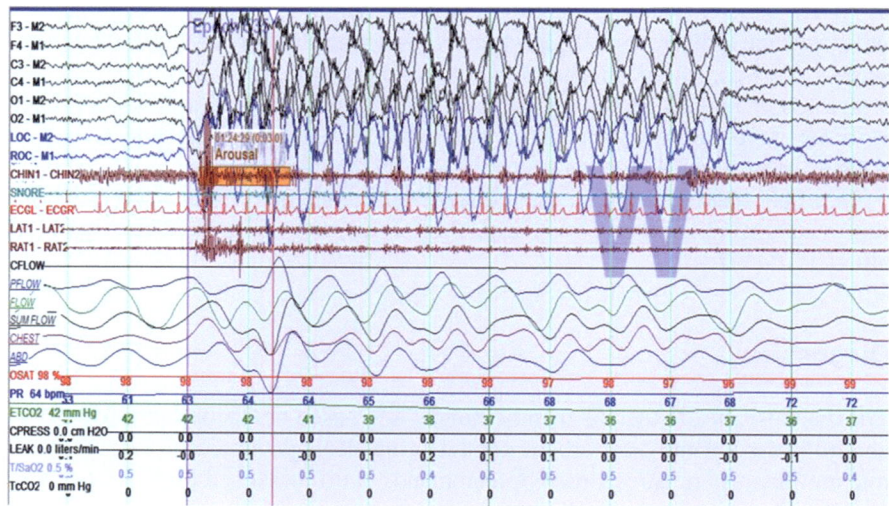

Fig. 12.1 Polysomnographic finding in patients with rhythmic movement disorder

occur during wakefulness and in every stage of sleep [15]. Rhythmic movements can occur during the sleep–wake transition or any sleep stage with various prevalences per sleep stage. It has been reported in 46% of NREM sleep, 30% of both NREM and REM sleep, and in 24% on only REM sleep [20], with the majority of events occurring in N2 [14]. The reporting of SRRMD occurring exclusively during REM sleep can raise the question of a possible relationship between SRRMD and REM behavior disorder (RBD) [21]. However, there is no current documentation of REM without atonia observed in patients with REM-related SRRMD and studies in patients with SRRMD exclusively during REM sleep have demonstrated normal atonia during REM [22].

Treatment Options

Currently, there are no evidence-based treatment guidelines or recommendations for the treatment of SRRMD in children or adults. In most cases, SRRMD does not result in significant concern or significant daytime symptoms and do not warrant treatment but in rare instances headbanging in particular has resulted in soft tissue, eye, and skull injuries, hemoglobinuria with acute renal failure, internal carotid artery dissection, and subdural hemorrhage [23].

Prescription Medication
Benzodiazepines have been used in children with various degrees of success. Oxazepam 10–20 mg at bedtime was used to treat body rocking and headbanging in an 8-year-old girl with symptoms since age 7 months [24]. In other reports, clonazepam has shown partial improvement in frequency or severity of movements, [3, 16] and in others there has been no improvement in symptoms [19].

Non-prescription
Melatonin was reported in a 8-year-old girl with ADHD and SRRMD showing improvement in frequency of RMD [25]. Hypnosis was reported successful in a case report of a 26-year-old woman with body-rocking since infant [26]. If comorbid OSA is present, CPAP may decrease the severity of SRRMD [27].

Restless Sleep Disorder

One of the most recent evidence of the rapidly evolving field of pediatric sleep is the recognition of RSD as a new pediatric disorder [28]. Children with RSD were initially identified because they typically present for evaluation of restless sleep characterized by their parents as moving frequently in their sleep, persistence of movements occurring all night, and importantly, associated with daytime symptoms of fatigue, sleepiness, or behavioral problems. When compared to children with restless legs syndrome (RLS) or normal controls, children with RSD do not show difficulty falling asleep, nor symptoms of RLS, or nocturnal awakenings [29]. A study assessing the presence, duration, and description of these movements during sleep using video-polysomnography confirmed that children with RSD moved in their sleep frequently and through the whole night [16]. In fact, these movements contributed to decreased total sleep time and increased brief arousals and awakenings, when compared to controls. When compared to children with RLS, as expected, children with RLS had increased indices of leg movements during sleep, not found in children with RSD. The International RLS Study Group (IRLSSG) formed a taskforce composed of 10 sleep experts to assess the literature and evaluate the evidence work that was summarized in the consensus diagnostic criteria for RSD [28]. Based on the medical literature and expert opinion, the task force established eight essential criteria for the diagnosis of RSD (Box 1) [28] which requires a video-polysomnogram performed in agreement with the American Academy of Sleep Medicine (AASM) Scoring Manual standard recommending at least 1 video frame per second [30], showing at least 5 movements per hour during sleep [29]. The current diagnostic criteria apply to children with RSD aged 6–18 years. The main reason younger children were not included is because of the need to rule out mimics. For instance, in the initial diagnostic publication, children unable to verbalize complaints of RLS or other discomfort were not included in the evaluation of RSD.

The prevalence of RSD in a sleep center-referred pediatric population has been estimated to be 7.7% [31].

Pathophysiology

The pathophysiology of RSD is still unknown, although we have postulated the following mechanisms which may act simultaneously and in synergy: iron deficiency, sleep instability, and increased sympathetic activation.

Iron Deficiency
Further studies have looked into potential mechanisms that explain RSD pathophysiology. Initially, all children with RSD were found to have mean ferritin levels of ~20 ng/dl. For other SRMD, such as RLS, non-anemic iron deficiency has been identified as a contributor to the symptoms, mainly by altering dopaminergic pathways involved in motor activity [32]. Children with RSD seem to improve clinically after iron supplementation, supporting this proposed mechanism [33].

Sleep Instability
Sleep instability was studied analyzing a physiological electroencephalographic pattern detected in polysomnography during NREM sleep which is a marker of NREM sleep instability, called cyclic alternating pattern (CAP) [34]. A recent study showed that children with RSD demonstrate abnormal CAP structure, suggesting alterations in sleep stability. These findings were characterized by lower percentage of A3 subtypes (arousals) than controls, shorter duration of the B phase of the CAP cycle, and shorter CAP cycle, indicating a faster than normal sleep instability, especially in slow-wave sleep. Another interesting finding was the fact that movements

in children with RSD occurred mainly during NCAP periods [35], interrupting them, confirming that the large muscle group movements seen in children with RSD during sleep are associated with significant sleep disruption. The movement clearly interrupts a period of stable nonREM sleep stage N2 with regular breathing is associated to an increase in heart rate and is followed by a prompt resumption of sleep.

Sympathetic Activation

Heart rate variability (HRV), a marker of sympathetic/parasympathetic balance, was analyzed in a group of children with RSD [36]. In normal sleep, the transition from wakefulness to sleep is accompanied by a switch from sympathetic to parasympathetic predominance manifested by slowing in heart rate and respiration [37]. HRV in children with RSD showed increased sympathetic activation during sleep compared to controls.

The natural course of RSD is not known. More research is needed to establish prognosis, confirm pathophysiology, and to identify RSD in younger children and adults.

Diagnosis

The International RLS Study Group formed a taskforce composed of 10 sleep experts to evaluate the evidence for RSD. Based on the medical literature and expert clinical experience, the task force found sufficient evidence for RSD and established eight essential criteria for diagnosis of RSD. The diagnosis of RSD includes clinical symptoms of restless sleep consisting of large movements during sleep, a duration of symptoms for more than 3 months and more than 3 nights a week and, importantly, video-polysomnographic evidence of five or more large body movements/hour. Children with RSD must have daytime impairment: either sleepiness, behavioral concerns, or cognitive deficits. Because all the research has been done in children aged 6–18, the taskforce established the criteria for children in this age range. Exclusion of mimics and other medical conditions need to be part of the initial evaluation and will be discussed in the next session.

Clinical and Laboratory Evaluation

The evaluation of a child with suspected RSD must include exclusion of mimics. Restless sleep is seen in 80% of children with restless legs syndrome (RLS) and 89% of children with periodic leg movements of sleep (PLMS) [38]. A review of the literature on restless sleep has identified a list of medical or sleep disorders that can present with restless sleep. Some of these disorders that need to be excluded are acute otitis media, asthma, pain, bruxism, among other disorders. Since restless sleep in children is commonly seen in association with other medical or sleep disorders, it is recommended to initiate the evaluation with a complete medical history and physical exam.

Synchronized video-polysomnography is required for the diagnosis of RSD. Its utility is two-fold, it can identify another sleep disorder such as obstructive sleep apnea or periodic limb movements disorder [39, 40], nocturnal seizures [41], eczema [42], among other disorders that can be associated with restless sleep. The AASM Scoring Manual [30] recommends an accuracy of at least 1 video frame per second when recording video PSG and recommends the use for video for the identification of sleep-associated movements, such as those seen in REM sleep behavior disorder (RBD), and rhythmic movement disorders (RMD). The diagnosis of RSD requires the demonstration of at least 5 movements per hour during vPSG [29].

Therapeutic Options

There are currently no recommendations for the treatment of RSD in children. Although the mechanism of the frequent movements during sleep is not fully elucidated, an association with iron deficiency has been identified. Iron is important in the synthesis of dopamine, a neurotransmitter that impacts motor activity [32]. RLS is the most studied sleep disorder in association with iron deficiency [43]. The International RLS Study Group has published guidelines recommending iron supplementation as first-line treatment for adults with RLS and supports the use of intravenous iron infusion for RLS [43]. Cho et al., suggested a dose-dependent improvement in symptoms after intravenous ferric carboxymaltose [44], with symptoms of RLS improving for up to 24 weeks after the infusion [45]. Expert consensus recommends iron supplementation in children when serum ferritin is less than 50 µg/L [43]. Iron supplementation in children with RLS and PLMD is usually the first-line treatment as well, accepted by most pediatric sleep physician [46].

Restless Legs Syndrome

According to the IRLSSG diagnostic criteria, RLS is a sleep-related movement disorder characterized by the need to move the legs, with symptoms appearing or worsening in the evening and during periods of rest, improving or disappearing with movement; according to the guidelines, symptoms in children must be able to be described in the child's own words [47].

The prevalence of the disorder in pediatric age is 2% in the general population and higher in sleep centers (>10%) [31]; this discrepancy could be attributable to a poor knowledge of this pathology, which consequently could be misdiagnosed or underdiagnosed, therefore it would be important to increase population studies on pediatric RLS, also involving pediatricians and local specialists (who should therefore be properly trained on this topic) and through the use of screening questionnaires for RLS specific to the pediatric age (which should therefore be appropriately drafted).

Pathophysiology

Pediatric RLS is highly familial, as demonstrated by the common occurrence in monozygotic twins and first-degree siblings. Both linkage and genotype analyses, as well as large cross-sectional studies show a positive family history in 87–90.9% of probands. Among the different genes associated to adult RLS (MEIS1, BTBD9, NAP2K5/SKOR1), only MEIS1 and SKOR1, supporting the sensory component of RLS, were found in association with the pediatric form. [48, 49]

At present, the iron deficiency-metabolic theory represents the most acclaimed explanation for the nature and distribution of RLS symptoms. Brain-iron deficiency (BID), due to impaired iron transport across the blood-brain barrier primarily affects the substantia nigra and, to a lesser extent, also the caudate, putamen and thalamus, with the activation of the hypoxic pathway. This, in turn, increases dopaminergic activity with subsequent post-synaptic downregulation. BID also increases glutamate and reduces adenosine, resulting in hyperarousal and sleep fragmentation unresponsive to dopaminergic treatment.

The circadian turnover of dopamine and iron metabolism also explains the circadian night distribution of symptoms and induces overcompensation via an adaptive post-synaptic mechanism [50].

Iron deficiency is currently recognized as an integral component of many neurodevelopmental and sleep disorders. In fact, it has been shown to strongly correlate with so-called hyperactive behaviors, including restless sleep disorder (RSD), ADHD, and ASD.

Clinical Features

The primary feature of RLS is the urge to move legs, with or without accompanying leg sensations. If sensations are present, they invariably involve the legs, though the arms and other body parts are sometimes affected. RLS is often misdiagnosed and is generally ignored by most pediatricians and general practitioners because of the mild and intermittent nature of the symptoms at younger ages or the inability of young children to characterize the sensations or discomfort in the lags. However, RLS is usually progressive and can cause significant functional impairment. Since RLS symptoms occur during bedtime, they are most likely to interfere with sleep onset and may be confused with bedtime resistance and limit setting-type behaviors. The majority of children with RLS (66%) report daytime leg discomfort differently from the typical increase during the evening or at night of the adults; this might be linked to the number of hours children spend sitting during the school day.

Diagnosis

The current ICSD-3 criteria for RLS are reported in Table 12.1 and several screening and diagnostic tools have been developed (seen in Tables 12.1 and 12.2).

Table 12.1 ICSD-3 criteria for restless legs syndrome

A. An urge to move the legs, usually accompanied by uncomfortable and unpleasant sensations in the legs:
  Begin or worsen during periods of rest or inactivity such as lying down or sitting
  Be partially or totally relieved by movement, such as walking or stretching, as least as long as the activity continues
  Occur exclusively or predominantly in the evening or night rather than during the day
B. The above features are not solely condition (e.g., leg cramps, positional discomfort, myalgia, venous stasis, leg oedema, arthritis, accounted for as symptoms of another medical or a behavioral habitual foot-tapping). For pediatric condition, there is the need to exclude mimics of RLS which are disorders
C. The symptoms of RLS cause concern, distress, sleep disturbance or impairment in mental, physical, social, occupational, educational, behavioral, or other important areas of functioning and cannot be better explained by other disorders, medication use, or substance use disorder

Table 12.2 Pediatric RLS diagnostic tools

Diagnostic interview
  Single question for RLS
  Pediatric Emory RLS diagnostic questionnaire
  The Restless Legs Syndrome questionnaire (RLSQ)
Severity scales
  International Restless Legs Scale (IRLS)
  Clinical Global Impressions Rating Scales (CGI)
  RLS-6 scale of restless legs syndrome/Willis-Ekbom disease
  Pediatric Restless Legs Syndrome Severity Scale (P-RLS-SS)
  Rating the four RLS diagnostic criteria
Contributing instrumental tools
  Video polysomnography (vPSG)
  Actimetry
  Suggested immobilization test (SIT)
  Suggested clinical immobilization test (SCIT)

The ICSD-3 states that "for children, the description of these symptoms should be in the child's own words." The interview questions should be phrased using words developmentally appropriate for the child. Language and cognitive development determine the applicability of the RLS diagnostic criteria, rather than age. As in adults, a significant impact on sleep, mood, cognition, and function is found. However, impairment is manifest more often in behavioral and educational domains. Periodic limb movement disorder may precede the diagnosis of RLS in some cases [1].

Differential Diagnosis

Differentiating pediatric RLS from other conditions, or mimics, can be complicated. Some of the common mimics of pediatric RLS are positional discomfort, sore leg muscles, ligament sprain/tendon strain, positional ischemia (numbness), dermatitis, bruises, growing pains, leg cramps, arthritis, peripheral neuropathy, radiculopathy, myelopathy, myopathy, fibromyalgia, and sickle cell disease.

Clinical and Laboratory Evaluation

Lab testing for anemia, iron profile, ferritin, and vitamin D levels should be part of the clinical evaluation. Physical and neurological examination of RLS children should prove within normal limits. A supportive objective criterion to strengthen the diagnosis of pediatric RLS may be obtained by PSG assessment of a PLMs index >5/h.

Consequences

RLS can have a very negative impact on the quality of life and sleep of the young patients (and their family members), as well as cause health problems or learning difficulties, considering the fundamental role of good sleep quality [51]. Furthermore, RLS is often associated with psychiatric problems such as anxiety, depression, or attention-deficit/hyperactivity disorder [52, 53], and in this case, the therapeutic management is often difficult; in fact, there are treatments used in the psychiatric field that can cause or aggravate RLS, or medications for RLS and psychiatric disorders can cause important side effects. However, specific therapeutic protocols for this pathology in pediatric age are lacking, in consideration of the few drugs approved by the FDA in children [54, 55];

RLS is often associated with the presence of PLMS (a condition that in adults has been correlated with changes in blood pressure and heart rate) [49], while very few studies have evaluated changes in blood pressure in children with RLS [56, 57], an important condition to be evaluated in consideration of the high prevalence of an high number of leg movements during sleep in children with RLS or other conditions [58, 59].

Treatment

Iron supplementation is the main treatment option for pediatric RLS [33]. Children with low ferritin levels (<35–50 ng/mL) receiving iron supplementation in the form of oral ferrous sulphate (3 mg/kg/day) improve their RLS symptoms after 3.8 months of therapy, showing an inverse correlation with their ferritin levels.

Table 12.3 Treatment options for RLS

Lifestyle	1. Exercise
	2. Diet
Interventions	1. Massage
	2. Heating pad
	3. Cooling
	4. Rubbing
	5. Sleep hygiene
Avoidance of	1. Antihistamines
	2. Antidepressants
	3. Caffeine
Iron supplementation	1. Oral ferrous sulfate
	2. Intravenous

Side effects include difficulty swallowing pills in children, constipation, or reflux. In celiac disease or whenever oral absorption is challenged or ferritin is very low, intravenous supplementation is advised, and a beneficial effect is reported with 1.2–6 mg/kg iron sucrose or 15 mg/kg up to 750 mg of Ferric Carboxymaltose (Table 12.3).

Periodic Leg Movement Disorder

PLMS are a polysomnographic finding (Fig. 12.2) usually corresponding to the extension of the first toe and dorsiflexion of the ankle that can also be accompanied by flexion of the knee and hip. PLMS are often bilateral but can be unilateral [60]. During polysomnography, electrodes attached to the skin over the tibialis anterior (TA) muscle are activated providing the typical muscle signal seen in the leg leads which allows to score these movements and also allows to identify if the PLMS are associated with an arousal or an awakening [61]. The number of PLMS per hour of sleep is referred to as the PLMS index (PLMI). The first challenge in the area of PLMD is that there are two current criteria for scoring PLMS which, especially in children, can provide very different results due to the fact that the old criteria by AASM [30] probably overestimate PLMS by considering periodic also nonperiodic leg activities [62]. The more recent criteria published in 2016 [63] seem to be more appropriate to describe the real degree of periodicity in children and should be used [58].

The second challenge involves the index used to support the potential diagnosis of PLMD. The AASM defines an abnormal PLMI as more than 5 in children and more than 15 in adults [1]. Recent research identifying PLMS across the lifespan has identified clear changes in the index and periodicity of leg movements across the different ages [59] and has demonstrated that an index of 2/h may be the best cut-off values for the PLMS index in children [58].

The clinical importance of PLMS is that they are most commonly seen in patients with RLS but can also be seen secondary to other conditions such as narcolepsy, REM sleep behavior disorder, or secondary to medication use. When PLMS are

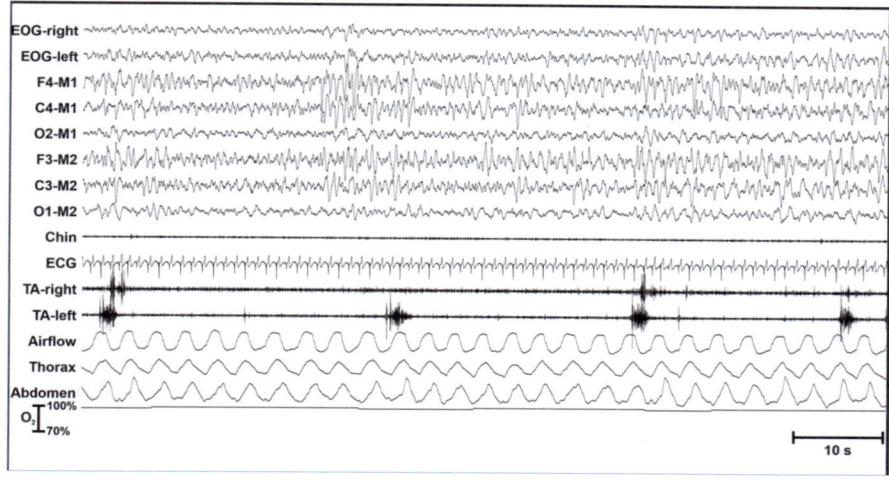

Fig. 12.2 Polysomnographic evidence of periodic limb movements of sleep

seen in isolation of other conditions and are thought to be responsible for daytime impairment in mental, behavioral, social, physical, or occupational area, PLMD) is diagnosed; an elevated PLMI by itself does not fit the criteria for a sleep disorder if it is not accompanied by other symptoms [1]. PLMD requires ruling out another primary sleep disorder that might explain the elevated PLMI, similarly PLMD is not diagnosed when PLMS are an incidental finding without any symptoms The prevalence and significance of PLMD in children is unknown mainly because of the variation in diagnostic parameters used in the pediatric literature, but some of the most accurate data may indicate a very low prevalence. For instance, DelRosso et al. [31] showed a single center prevalence of 0.3%, and Kirk and Bohn [64] found that a suspicion of PLMD based on PLMI >5 and no other comorbidity, was 1.2%. One of the reasons for the challenging identification of PLMD is also presented by the fact that children with elevated PLMI often have many comorbidities and are on various medications [65].

Comparison between RSD, RLS, and PLMD

	RSD	RLS	PLMD
Clinical presentation	Restless sleep Daytime impairment	Urge to move legs Sleep-onset insomnia	Periodic leg movements of sleep Daytime impairment
Diagnosis	Clinical + PSG	Clinical	Clinical + PSG
PSG findings	Body movement index >5 per hour	May or may not have elevated PLMI	PLMI >5
Pathophysiology	Sleep instability Iron deficiency Sympathetic activation	Dopamine dysfunction Iron deficiency	Unknown

Bruxism

Pediatric bruxism, commonly known as teeth grinding, is a condition characterized by the involuntary clenching or grinding of teeth, typically during sleep. This condition can affect children of all ages and can have significant implications for their dental health, sleep quality, and overall well-being [66]. Understanding the pathophysiology, diagnosis, differential diagnosis, consequences, and treatment of pediatric bruxism is essential for sleep specialists.

Pathophysiology

The exact pathophysiology of pediatric bruxism is not fully understood, but it is believed to be multifactorial, involving a combination of genetic, psychological, and physiological factors. Several theories have been proposed to explain the mechanisms underlying bruxism:

Central Nervous System (CNS) Involvement
It is hypothesized that bruxism may be related to areas in the central nervous system, particularly in the areas responsible for motor control and arousal. Increased activity in the brainstem and basal ganglia during sleep may trigger the involuntary muscle contractions that lead to teeth grinding [67].

Genetic Predisposition
There is evidence to suggest that bruxism may have a genetic component. Studies have shown that children with a family history of bruxism are more likely to develop the condition themselves [67].

Psychological Factors
Stress, anxiety, and emotional disturbances are commonly associated with bruxism. Children experiencing high levels of stress or anxiety may be more prone to teeth grinding as a coping mechanism [67].

Sleep Disorders
Bruxism is often linked to other sleep disorders, such as obstructive sleep apnea (OSA) and restless legs syndrome (RLS). These conditions can disrupt normal sleep patterns and increase the likelihood of bruxism episodes.

Dental Factors
Malocclusion, or misalignment of the teeth, can contribute to bruxism. Children with dental issues such as overbites, underbites, or crowded teeth may be more susceptible to grinding their teeth.

Diagnosis

Diagnosing pediatric bruxism involves a comprehensive evaluation that includes a detailed medical and dental history, clinical examination, and sometimes additional diagnostic tests. Key steps in the diagnostic process include gathering information about the child's medical history, family history of bruxism, and any psychological or emotional factors that may be contributing to the condition. Parents may also be asked about the child's sleep patterns and any observed teeth grinding during sleep [68].

Clinical and Laboratory Evaluation

A thorough examination of the child's teeth, jaw, and oral structures is essential. Signs of bruxism may include worn tooth enamel, flattened or chipped teeth, and tenderness in the jaw muscles. Parents may be asked to complete questionnaires or keep sleep diaries to document the frequency and severity of the child's bruxism episodes [68]. This information can help in assessing the impact of bruxism on the child's sleep and daily functioning, sleep patterns, and detect any associated sleep disorders. Polysomnography can provide valuable insights into the frequency and duration of bruxism episodes and their relationship to other sleep disturbances like obstructive sleep apnea.

Differential Diagnosis

Differentiating pediatric bruxism from other conditions with similar symptoms is crucial for accurate diagnosis and effective treatment. Conditions that may mimic or coexist with bruxism include Temporomandibular Joint Disorders (TMD) can cause jaw pain and dysfunction, which may be mistaken for bruxism. A thorough examination of the temporomandibular joint (TMJ) can help distinguish between these conditions.

Misalignment of the teeth can contribute to bruxism, but it can also cause other dental issues. Identifying and addressing malocclusion is important in the management of bruxism.

Children with OSA may exhibit bruxism as a secondary symptom. Polysomnography can help diagnose OSA and determine its relationship to bruxism.

Nocturnal seizures can sometimes present with teeth grinding or jaw clenching. A detailed medical history and, if necessary, an electroencephalogram (EEG) can help differentiate between bruxism and seizure activity.

Consequences

Pediatric bruxism can have several adverse consequences, affecting both dental health and overall well-being. Chronic teeth grinding can lead to significant dental wear, including enamel erosion, tooth fractures, and increased tooth sensitivity. In severe cases, it may result in damage to dental restorations and the need for extensive dental treatment.

Bruxism can cause muscle fatigue, pain, and tenderness in the jaw muscles. Over time, this can lead to temporomandibular joint (TMJ) disorders, characterized by pain, clicking, and limited jaw movement.

Frequent bruxism episodes can disrupt sleep architecture, leading to fragmented and non-restorative sleep. This can result in daytime sleepiness, irritability, and impaired cognitive and behavioral functioning.

The stress and anxiety associated with bruxism can have a negative impact on a child's emotional well-being. Children may experience increased anxiety, mood disturbances, and difficulties in coping with daily stressors.

Treatment

Stress management techniques, relaxation exercises, and cognitive-behavioral therapy (CBT) can help reduce the psychological factors contributing to bruxism. Teaching children coping mechanisms and relaxation techniques can be beneficial [66].

Custom-made dental appliances, such as mouthguards or splints, can protect the teeth from further damage and reduce the impact of grinding. These devices are typically worn at night and can help alleviate jaw pain and prevent dental wear [66].

In some cases, medications may be prescribed to manage bruxism. Muscle relaxants, anti-anxiety medications, or medications to address underlying sleep disorders may be considered. However, pharmacological treatments are generally used with caution in children.

If bruxism is associated with other sleep disorders, such as obstructive sleep apnea, treating the underlying sleep disorder is essential. Continuous positive airway pressure (CPAP) therapy or other interventions may be recommended to improve sleep quality.

Regular visits to the dentist are crucial for monitoring the child's dental health and addressing any issues related to bruxism. Early intervention can prevent long-term dental damage and ensure appropriate management.

Conclusion

Sleep-related movement disorders in children represent a diverse and complex group of conditions that can profoundly affect sleep quality and overall health. These disorders, including sleep-related rhythmic movement disorders, restless sleep disorder, restless legs syndrome, periodic limb movement disorder (PLMD), and bruxism, each have unique characteristics and challenges. Understanding these disorders is crucial for healthcare providers, educators, and parents to ensure that affected children receive appropriate care and support.

Early recognition and accurate diagnosis of sleep-related movement disorders are essential. Timely intervention can prevent the progression of symptoms and mitigate the impact on a child's development, comorbidities, and daily functioning. Sleep specialists should be vigilant in identifying signs and symptoms, utilizing diagnostic tools, such as polysomnography, when necessary, to confirm the presence of these disorders.

Effective management of sleep-related movement disorders requires a comprehensive approach that addresses the underlying causes, alleviates symptoms, and improves sleep quality. Treating brain-iron deficiency in children with low ferritin is a harmless and beneficial approach despite no FDA-approved treatments exist for the management of these disorders in children.

Pearls/Take-Home Points

- Sleep-related movement disorders can significantly impact sleep quality and overall health, particularly in pediatric patients.
- Restless Sleep Disorder (RSD) is a newly defined entity that requires video-polysomnography for diagnosis and is associated with iron deficiency, sleep instability, and sympathetic activation.
- Restless Legs Syndrome (RLS) in children is often familial, related to iron deficiency, and misdiagnosed due to atypical symptom presentation.
- Periodic Limb Movement Disorder (PLMD) should be carefully distinguished from incidental periodic limb movements, as only symptomatic cases require treatment.
- Treatment strategies include iron supplementation, behavioral interventions, and in some cases, pharmacologic therapy tailored to individual patient needs.

References

1. American Academy of Sleep Medicine. International classification of sleep disorders 3rd edition Darien. American Academy of Sleep Medicine; 2014.
2. Su C, Miao J, Liu Y, et al. Multiple forms of rhythmic movements in an adolescent boy with rhythmic movement disorder. Clin Neurol Neurosurg. 2009;111(10):896–9.
3. Mayer G, Wilde-Frenz J, Kurella B. Sleep related rhythmic movement disorder revisited. J Sleep Res. 2007;16(1):110–6.
4. Happe S, Ludemann P, Ringelstein EB. Persistence of rhythmic movement disorder beyond childhood: a videotape demonstration. Mov Disord. 2000;15(6):1296–8.
5. Sallustro F, Atwell CW. Body rocking, head banging, and head rolling in normal children. J Pediatr. 1978;93(4):704–8.

6. Terzano MG, Parrino L, Sherieri A, et al. Atlas, rules, and recording techniques for the scoring of cyclic alternating pattern (CAP) in human sleep. Sleep Med. 2001;2(6):537–53.
7. Terzano MG, Parrino L. Origin and significance of the Cyclic Alternating Pattern (CAP). REVIEW ARTICLE. Sleep Med Rev. 2000;4(1):101–23.
8. Parrino L, Smerieri A, Spaggiari MC, Terzano MG. Cyclic alternating pattern (CAP) and epilepsy during sleep: how a physiological rhythm modulates a pathological event. Clin Neurophysiol. 2000;111(Suppl 2):S39–46.
9. Manni R, Terzaghi M. Rhythmic movements during sleep: a physiological and pathological profile. Neurol Sci. 2005;26(Suppl 3):s181–5.
10. Manni R, Terzaghi M, Sartori I, Veggiotti P, Parrino L. Rhythmic movement disorder and cyclic alternating pattern during sleep: a video-polysomnographic study in a 9-year-old boy. Mov Disord. 2004;19(10):1186–90.
11. Freund HJ, Hefter H. The role of basal ganglia in rhythmic movement. Adv Neurol. 1993;60:88–92.
12. Grillner S, Wallen P. Central pattern generators for locomotion, with special reference to vertebrates. Annu Rev Neurosci. 1985;8:233–61.
13. Attarian H, Ward N, Schuman C. Case report: a multigenerational family with persistent sleep related rhythmic movement disorder (RMD) and insomnia. J Clin Sleep Med. 2009;5(6):571–2.
14. Dyken ME, Lin-Dyken DC, Yamada T. Diagnosing rhythmic movement disorder with video-polysomnography. Pediatr Neurol. 1997;16(1):37–41.
15. Stepanova I, Nevsimalova S, Hanusova J. Rhythmic movement disorder in sleep persisting into childhood and adulthood. Sleep. 2005;28(7):851–7.
16. Hashizume Y, Yoshijima H, Uchimura N, Maeda H. Case of head banging that continued to adolescence. Psychiatry Clin Neurosci. 2002;56(3):255–6.
17. Chiaro G, Maestri M, Riccardi S, et al. Sleep-related rhythmic movement disorder and obstructive sleep apnea in five adult patients. J Clin Sleep Med. 2017;13(10):1213–7.
18. Yeh SB, Schenck CH. Atypical headbanging presentation of idiopathic sleep related rhythmic movement disorder: three cases with video-polysomnographic documentation. J Clin Sleep Med. 2012;8(4):403–11.
19. Lee SK. A case with dopamine-antagonist responsive repetitive head punching as rhythmic movement disorder during sleep. J Epilepsy Res. 2013;3(2):74–5.
20. Hoban TF. Rhythmic movement disorder in children. CNS Spectr. 2003;8(2):135–8.
21. Kempenaers C, Bouillon E, Mendlewicz J. A rhythmic movement disorder in REM sleep: a case report. Sleep. 1994;17(3):274–9.
22. Kohyama J, Matsukura F, Kimura K, Tachibana N. Rhythmic movement disorder: polysomnographic study and summary of reported cases. Brain Dev. 2002;24(1):33–8.
23. Mackenzie JM. "Headbanging" and fatal subdural haemorrhage. Lancet. 1991;338(8780):1457–8.
24. Walsh JK, Kramer M, Skinner JE. A case report of jactatio capitis nocturna. Am J Psychiatry. 1981;138(4):524–6.
25. Metin O, Eynalli-Gok E, Kuygun-Karci C, Yolga-Tahiroglu A. The effectiveness of melatonin in head banging: a case report. Sleep Sci. 2019;12(1):53–6.
26. Rosenberg C. Elimination of a rhythmic movement disorder with hypnosis–a case report. Sleep. 1995;18(7):608–9.
27. Chirakalwasan N, Hassan F, Kaplish N, Fetterolf J, Chervin RD. Near resolution of sleep related rhythmic movement disorder after CPAP for OSA. Sleep Med. 2009;10(4):497–500.
28. DelRosso LM, Ferri R, Allen RP, et al. Consensus diagnostic criteria for a newly defined pediatric sleep disorder: restless sleep disorder (RSD). Sleep Med. 2020;75:335–40.
29. DelRosso LM, Bruni O, Ferri R. Restless sleep disorder in children: a pilot study on a tentative new diagnostic category. Sleep. 2018;41(8) https://doi.org/10.1093/sleep/zsy102.
30. Berry RB, Brooks R, Gramaldo CE, et al. For the American Academy of sleep Medicine. In: The AASM manual for the scoring of sleep and associated events, vol. 2.4. Darien: AASM; 2017.

31. DelRosso LM, Ferri R. The prevalence of restless sleep disorder among a clinical sample of children and adolescents referred to a sleep centre. J Sleep Res. 2019;28:e12870.
32. Angulo-Barroso RM, Peirano P, Algarin C, Kaciroti N, Lozoff B. Motor activity and intra-individual variability according to sleep-wake states in preschool-aged children with iron-deficiency anemia in infancy. Early Hum Dev. 2013;89(12):1025–31.
33. DelRosso LM, Yi T, Chan JHM, Wrede JE, Lockhart CT, Ferri R. Determinants of ferritin response to oral iron supplementation in children with sleep movement disorders. Sleep. 2020;43(3):zsz234.
34. DelRosso LM, Jackson CV, Trotter K, Bruni O, Ferri R. Video-polysomnographic characterization of sleep movements in children with restless sleep disorder. Sleep. 2019;42:zsy269.
35. Parrino L, Smerieri A, Rossi M, Terzano MG. Relationship of slow and rapid EEG components of CAP to ASDA arousals in normal sleep. Sleep. 2001;24(8):881–5.
36. Ferri R, Parrino L, Smerieri A, et al. Cyclic alternating pattern and spectral analysis of heart rate variability during normal sleep. J Sleep Res. 2000;9(1):13–8.
37. Snyder F, Hobson JA, Morrison DF, Goldfrank F. Changes in respiration, heart rate, and systolic blood pressure in human sleep. J Appl Physiol. 1964;19:417–22.
38. Picchietti DL, Rajendran RR, Wilson MP, Picchietti MA. Pediatric restless legs syndrome and periodic limb movement disorder: parent-child pairs. Sleep Med. 2009;10(8):925–31.
39. Medicine AAoS. International classification of sleep disorders. 3rd ed. Darien: American Academy Of Sleep Medicine; 2014.
40. Walter LM, Nixon GM, Davey MJ, et al. Differential effects of sleep disordered breathing on polysomnographic characteristics in preschool and school aged children. Sleep Med. 2012;13(7):810–5.
41. Wang X, Marcuse LV, Jin L, et al. Sleep-related hypermotor epilepsy activated by rapid eye movement sleep. Epileptic Disord. 2018;20(1):65–9.
42. Camfferman D, Kennedy JD, Gold M, Simpson C, Lushington K. Sleep and neurocognitive functioning in children with eczema. Int J Psychophysiol. 2013;89(2):265–72.
43. Allen RP, Picchietti DL, Auerbach M, et al. Evidence-based and consensus clinical practice guidelines for the iron treatment of restless legs syndrome/Willis-Ekbom disease in adults and children: an IRLSSG task force report. Sleep Med. 2018;41:27–44.
44. Cho YW, Allen RP, Earley CJ. Efficacy of ferric carboxymaltose (FCM) 500 mg dose for the treatment of Restless Legs Syndrome. Sleep Med. 2018;42:7–12.
45. Allen RP, Adler CH, Du W, Butcher A, Bregman DB, Earley CJ. Clinical efficacy and safety of IV ferric carboxymaltose (FCM) treatment of RLS: a multi-centred, placebo-controlled preliminary clinical trial. Sleep Med. 2011;12(9):906–13.
46. Picchietti DL. Restless legs syndrome/Willis-Ekbom disease and periodic limb movement disorder in children. In: Basow DS, editor. UpToDate. Waltham: UpToDate; 2020.
47. Allen RP, Picchietti DL, Garcia-Borreguero D, et al. Restless legs syndrome/Willis-Ekbom disease diagnostic criteria: updated International Restless Legs Syndrome Study Group (IRLSSG) consensus criteria–history, rationale, description, and significance. Sleep Med. 2014;15(8):860–73.
48. Picchietti MA, Picchietti DL. Advances in pediatric restless legs syndrome: iron, genetics, diagnosis and treatment. Sleep Med. 2010;11(7):643–51.
49. Trenkwalder C, Allen R, Hogl B, et al. Comorbidities, treatment, and pathophysiology in restless legs syndrome. Lancet Neurol. 2018;17(11):994–1005.
50. Rizzo G, Plazzi G. Neuroimaging applications in restless legs syndrome. Int Rev Neurobiol. 2018;143:31–64.
51. Licis A. Sleep disorders: assessment and treatment in preschool-aged children. Child Adolesc Psychiatr Clin N Am. 2017;26(3):587–95.
52. Pullen SJ, Wall CA, Angstman ER, Munitz GE, Kotagal S. Psychiatric comorbidity in children and adolescents with restless legs syndrome: a retrospective study. J Clin Sleep Med. 2011;7(6):587–96.
53. Angriman M, Cortese S, Bruni O. Somatic and neuropsychiatric comorbidities in pediatric restless legs syndrome: a systematic review of the literature. Sleep Med Rev. 2017;34:34–45.

54. DelRosso L, Bruni O. Treatment of pediatric restless legs syndrome. Adv Pharmacol. 2019;84:237–53.
55. DelRosso LM, Mogavero MP, Baroni A, Bruni O, Ferri R. Restless legs syndrome in children and adolescents. Child Adolesc Psychiatr Clin N Am. 2021;30:143.
56. Walter LM, Foster AM, Patterson RR, et al. Cardiovascular variability during periodic leg movements in sleep in children. Sleep. 2009;32(8):1093–9.
57. DelRosso LM, Mogavero MP, Ferri R. Effect of sleep disorders on blood pressure and hypertension in children. Curr Hypertens Rep. 2020;22(11):88.
58. Ferri R, DelRosso LM, Arico D, et al. Leg movement activity during sleep in school-age children and adolescents: a detailed study in normal controls and participants with restless legs syndrome and narcolepsy type 1. Sleep. 2018;41(4):zsy010.
59. Ferri R, DelRosso LM, Silvani A, et al. Peculiar lifespan changes of periodic leg movements during sleep in restless legs syndrome. J Sleep Res. 2020;29(3):e12896.
60. Symonds CP. Nocturnal myoclonus. J Neurol Neurosurg Psychiatry. 1953;16(3):166–71.
61. Sforza E, Nicolas A, Lavigne G, Gosselin A, Petit D, Montplaisir J. EEG and cardiac activation during periodic leg movements in sleep: support for a hierarchy of arousal responses. Neurology. 1999;52(4):786–91.
62. Ferri R, Rundo F, Zucconi M, et al. Putting the periodicity back into the periodic leg movement index: an alternative data-driven algorithm for the computation of this index during sleep and wakefulness. Sleep Med. 2015;16(10):1229–35.
63. Ferri R, Fulda S, Allen RP, et al. World Association of Sleep Medicine (WASM) 2016 standards for recording and scoring leg movements in polysomnograms developed by a joint task force from the International and the European Restless Legs Syndrome Study Groups (IRLSSG and EURLSSG). Sleep Med. 2016;26:86–95.
64. Kirk VG, Bohn S. Periodic limb movements in children: prevalence in a referred population. Sleep. 2004;27(2):313–5.
65. Delrosso LM, Lockhart C, Wrede JE, et al. Comorbidities in children with elevated periodic limb movement index during sleep. Sleep. 2020;43(2):zsz221.
66. Matusz K, Maciejewska-Szaniec Z, Gredes T, et al. Common therapeutic approaches in sleep and awake bruxism – an overview. Neurol Neurochir Pol. 2022;56(6):455–63.
67. Lavigne GJ, Khoury S, Abe S, Yamaguchi T, Raphael K. Bruxism physiology and pathology: an overview for clinicians. J Oral Rehabil. 2008;35(7):476–94.
68. Lobbezoo F, Ahlberg J, Raphael KG, et al. International consensus on the assessment of bruxism: report of a work in progress. J Oral Rehabil. 2018;45(11):837–44.

Excessive Daytime Sleepiness in Pediatric Age

13

Aysegul Karaca, Maria Rodriguez, Hadia Tahir, and Syed Kamal Naqvi

EDS

A typical patient with excessive daytime sleepiness (EDS) presents with daytime somnolence, fatigue, and difficulty staying alert [1]. The propensity to fall asleep is result of chronic sleep deprivation resulting in loss of daytime function. EDS is reported by 10–25% of the general population [2]. Symptoms of EDS are mood changes, difficulty concentrating, napping during school hours, feeling lethargic. In the pediatric population, EDS can be caused by different factors such as chronic insufficient sleep, insomnia, obstructive sleep apnea, narcolepsy, or mental health disorders, such as anxiety and depression [3]. Paradoxical symptoms in younger children may manifest as inattentiveness and hyperactivity and impulsiveness in older children [3]. In adolescents, EDS can be seen as apathy, poor language, flat affectivity, memory disorder, inflexible thinking, and lack of planning. Consequences of EDS may lead to impaired academic performance, mental and physical.

Daytime fatigue is a common complaint characterized by persistent lack of energy and tiredness not relieved by rest. Often individuals are impacted mentally and physically that can impact an individual's quality of life. It is difficult to distinguish between sleepiness and fatigue therefore symptoms can overlap [4]. See Fig. 13.1 for illustration of EDS symptoms.

Fig. 13.1 Illustration of symptoms of daytime sleepiness

Incidence

Studies have shown that daytime sleepiness increases with age. Excessive daytime sleepiness is a global phenomenon affecting 40% of children and adolescents [5]. The prevalence of excessive daytime sleepiness (PDSS >18) increased from 19.8% at prepuberty to 47.2% at post-puberty [3]. Stress, electronic use, short sleep duration, and sleep disturbances are associated factors causing excessive daytime sleepiness adolescents [3].

Examination/Mental Status Exam

It is important to take a complete sleep history and physical examination and perform specific testing to exclude other comorbid conditions of EDS. Incorporating a sleep questionnaire during routine visits is valuable. Sleep questionnaires or screen tools can provide the clinician information including difficulty with initiating sleep or maintaining sleep, restless sleep, parasomnias, night-time awakenings, snoring and sleep-disordered breathing can result in sleep fragmentation [6].

A general physical examination provides the clinician valuable insight on the patient's level of somnolence. Evaluation includes the child's level of alertness, repetitive yawning, droopy eyelids, flat affect, hyperactivity, and irritability which may indicate EDS. Patients may also exhibit daytime mouth breathing associated with nasal congestion and noisy breathing. Obese patients are at higher risk of sleep-disordered breathing leading to poor quality of sleep [6].

Patients body habitus is noteworthy as obese patients are at higher risk of obstructive sleep apnea OSA. Studies have shown OSA is prevalent in 4% of adolescents (16–19 years of age) and most of them without previous history of CSA or habitual snoring during mid-childhood [7]. Patients' quality of life can be impacted negatively with the presence of OSA which can be associated with EDS, fatigue, and neurocognitive dysfunction [2].

It is important to evaluate the patient's neurocognitive status. Sleep fragmentation can lead to impaired daytime function affecting mood (irritability and anxiety), behavior, attention, concentration, memory, coordination, problem-solving, and decision-making [6]. School-age children and adolescents are at risk of academic decline or failure.

An essential part of the physical examination are head, eyes, ears, neck, and throat (HEENT). Adenotonsillar hypertrophy and high Mallampati score are possible risk factors for sleep disordered breathing [6]. A term used for individuals with characteristics of enlarged adenoids is adenoidal facie. Traits include long face, open mouth, muffled voice, dental problems, and upper jaw underdevelopment. Adenoid hypertrophy is a condition associated with sleep-disordered breathing, sinus infections, and seasonal allergies. Other findings on exam suggesting sleep disturbances are allergic shiners, inflammation of the nasal passageway, and nasal congestion [6].

Examine mandible position in relation to maxilla and other facial abnormalities. Sleep-disordered breathing can be associated with retrognathia and micrognathia conditions. Craniofacial abnormalities can be associated with Down Syndrome, Pierre Robin, and cleft and palate abnormalities. Hypotonia can also contribute to upper airway collapse causing worsening of OSA [6].

Evaluation

Diagnostic tools that can be utilized to evaluate sleep disturbances are sleep diaries or logs, actigraphy, or overnight polysomnography [8]. A sleep diary or log are helpful to collect data on bedtime, sleep onset, night-time awakening's (number and duration), morning waking, and naps for a two-week duration.

The Epworth Sleepiness Scale measures daytime sleepiness [9]. The questionnaire consists of 8 questions rating their likelihood of falling asleep in normal day-to-day activities. Scale from 0 to 3 is utilized [10].

- **0**: Would never doze
- **1**: Slight chance of dozing
- **2**: Moderate chance of dozing
- **3**: High chance of dozing

Questions on the ESS

The following typical scenarios are provided in the ESS:

1. Sitting and reading
2. Watching TV
3. Sitting inactive in a public place (e.g., theater or meeting)
4. As a passenger in a car for an hour without a break
5. Lying down to rest in the afternoon when circumstances permit
6. Sitting and talking to someone
7. Sitting quietly after lunch (without alcohol)
8. In a car, while stopped for a few minutes in traffic

Interpretation

- A total score ranges from 0 to 24. Higher scores indicate greater daytime sleepiness (Table 13.1).

Table 13.1 Epworth sleepiness scale

Score	Interpretation
0–5	Low level of normal daytime sleepiness
6–10	Higher level of normal daytime sleepiness
11–12	Mild excessive daytime sleepiness
13–15	Moderate excessive daytime sleepiness
16–24	Severe excessive daytime sleepiness

The Standford Sleepiness Scale (SSS) is a self-reported questionnaire measuring sleepiness at any given moment. It is a research tool to assess a person's subjective level of sleepiness. SSS is a scale ranging from 1 to 7. One of seven statements is chosen that best describes an individual's level of sleepiness [10].

- 1 = feeling active, vital, alert, wide awake
- 2 = functioning at a high level, not at peak, able to concentrate
- 3 = relaxed, awake, not at full alertness, responsive
- 4 = a little foggy, not at peak, let down
- 5 = fogginess, losing interest in remaining awake, slowed
- 6 = sleepiness, prefer to be lying down, fighting sleep, woozy
- 7 = almost in reverie, sleep onset soon, losing struggle to remain awake

Individuals who experience excessive daytime sleepiness may select statements fourth through seventh when they should feel alert.

A relationship exists between excessive daytime sleepiness and the severity of depression. Depression is one of the most common psychiatric illnesses in teenagers, including OSA with EDS [11]. Screening for depression is warranted for patients with sleep-disordered breathing who can commonly demonstrate depressive symptoms and impaired daytime performance. Yearly depression screening is recommended for children and adolescents. Patient Health Questionnaire (PHQ-9) is a modified version of the adult PHQ-9. The screening tool consists of 9 questions based on the DSM-5 criteria for depression [2].

Actigraphy is a lightweight, compact wristwatch which records an individual's sleep pattern over a 7-to-14-day period 6. Child-specific algorithms on the watch sense sleep and wake episodes. The individual is in their home environment. The American Academy of Sleep Medicine (AASM) states that the use of actigraphy in normal children and special pediatric populations is indicated for the assessment of sleep patterns and response to treatment [12].

An overnight sleep study, or polysomnography, is used to diagnose sleep disorders such as central apnea, obstructive sleep apnea, or periodic limb movements. During the overnight sleep study, sensors are placed to monitor brain activity (EEG), heart rate, respiratory rate, pulse oximetry, noninvasive CO_2 monitoring, sleep architecture, and lower extremity muscle movement [6].

The Multiple Sleep Latency Test (MSLT) measures the propensity of how quickly an individual can fall asleep. It is part of the routine evaluation of patients who have EDS and may also be helpful in evaluation patients with suspected idiopathic hypersomnia or narcolepsy. Sleep study takes place the night prior to the MSLT to evaluate patients sleep and look for sleep disturbance leading to poor quality of sleep. Naps are offered at 2-h intervals until the patient had five opportunities to nap. The mean latency sleep is measured and presence or absence of sleep onset REM (SOREM) which occurs in the first 15 min of falling asleep [8].

Other factors that can contribute to EDS and fatigue include anemia, abnormal thyroid levels, and drug use. Studies have shown iron deficiency can lead to restless sleep. Ferritin is a protein that stores iron and measured by a blood test [6]. Ferritin levels below 50 mcg are treated with iron supplements over a 3-month period [6]. Re-evaluation of symptoms and ferritin level are completed at follow up.

Exam

General appearance: sleepy obese adolescent with allergic shiners.
Oral Cavity:
Oropharyngeal space: small
Tongue size: large
Palate: normal
Tonsils size: 3
Mallampatti Grade: 1-open
Sleep diary: A 2-week sleep diary was kept recording sleep hours and wake up times. Sleep diary indicates variable bedtime during the weekdays with frequent awakenings during the night. On weekends bedtime consistently after 2 am. Rise time consistently at 630 am for school on weekdays. Weekends wake up time after 1 pm. No nap time recorded.

Polysomnogram

On room air: The study demonstrated profound obstructive sleep apnea (oAHI 68/h) with mild hypoxemia and no hypercapnia. Loud snoring was audible.

There was a REM sleep-related worsening noted in obstructive events.

Sleep was fragmented with frequent arousals and awakenings. REM sleep was reduced.

Limited EKG and EEG were within normal limits.

CPAP was tried due to severity of sleep-disordered breathing and was found to be helpful in improving obstructive events, hypoxemia and hypercapnia.

During titration CPAP was tried with pressure ranging from 4 to 9 cwp with optimal pressure being 8 cwp with oAHI 2/hr., SpO2 above 95% and pCO2 < 48 mmHg.

Improved sleep efficiency but supine REM sleep was not achieved. Patient did tolerate PAP well.

No significant periodic limb movements were present.
Actigraphy report (Fig. 13.2).

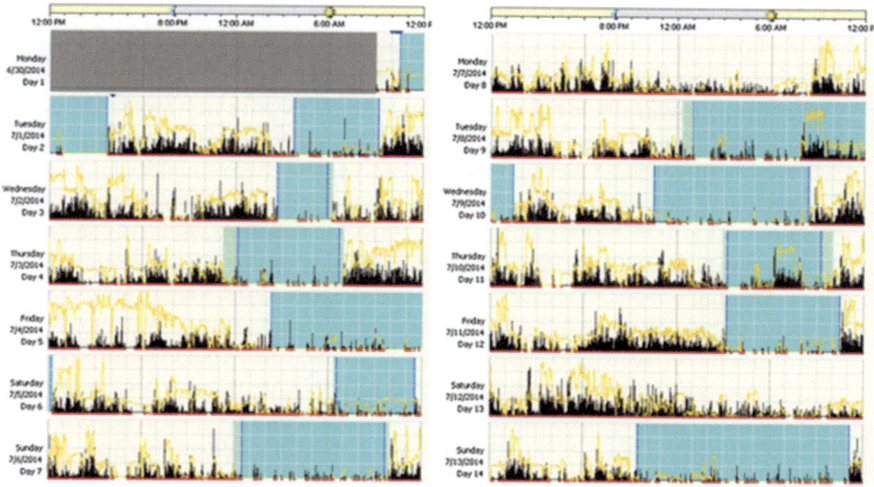

Fig. 13.2 Actigraphy report for clinical case indicating delayed sleep phase

Differential Diagnosis (Comorbid Conditions)

Daytime excessive sleepiness in pediatric age could be associated with genetic and metabolic disorders, as well as behavioral patterns and mental health.

1. Metabolic Causes:
    1.1 Hypothyroidism: One of the very well-known metabolic disorders that is associated with excessive daytime sleepiness is hypothyroidism. Studies have shown that thyroid hormone plays a key role in modulation of circadian rhythm [13]. Hypothyroid babies have decreased REMS. Children whose congenital hypothyroidism is not corrected by thyroid medication are known to have a reduction of slow wave sleep and impairment of sleep spindle development [13]. In conclusion, hypothyroidism affects sleep structure and decreases the quality of sleep that leads to excessive daytime sleepiness. Adequate management of hypothyroidism is warranted to mitigate EDS symptoms.
    1.2 Diabetes: There is bidirectional relation between sleep and blood glucose, poor sleep affects glucose management, as well as uncontrolled glucose affect sleep [14]. Interruption of sleep due to blood sugar checks, insulin administration leads decreased sleep quality. Silva et al. reported increased HgA1c level ($\geq 7.5\%$) and short duration after diagnosis of Type 1 DM is associated with EDS [14]. Studies support better glycemic control would decrease EDS in children with Type 1 DM.

1.3 Iron deficiency and restless leg syndrome: Prevalence of restless leg is 2–4% in children and adolescents [15]. Diagnosis of RLS is clinical and "urge for moving legs" is the key component of the clinical picture. Urge for leg movement worsens in the evening and improves with movement shouldn't be secondary to another condition [15]. Single question "When you try to relax in the evening or sleep at night, do you ever have unpleasant, restless feelings in your legs that can be relieved by walking or movement" can detect RLS with 100% sensitivity and 96.8% specificity with a negative predictive value of 100%. Polysomnographic study can be used to support RLS, interestingly isolated leg movements can present earlier than the RLS symptom onset [15]. EDS and problems at school are the most prevalent daytime symptoms of RLS [15], followed by disrupted sleep, behavioral problems such as irritability, aggression or hyperactivity, attention deficit, and mood changes [15]. 85% of children with RLS are found to have a Ferritin level less than 50 ng/mL. Although there is not enough evidence for iron supplementation in RLS, it is commonly recommended for RLS for children with Ferritin level less than 50 ng/mL besides nonpharmacological approach including sleep hygiene and lifestyle modification with healthy diet and adequate physical activity.
2. Central disorder of hypersomnolence (CDH) is a cluster of neurological sleep disorders including narcolepsy, idiopathic hypersomnia, and Kleine–Levin syndrome [16].
   2.1 Narcolepsy is a complex sleep disorder and is the most common cause of EDS in pediatric age, which can occur with or without cataplexy [17]. Narcolepsy with cataplexy is classified as type 1, whereas narcolepsy without cataplexy is identified as Type 2. Prevalence is 0.025–0.05% in the population with a peak at ages 10–19 [17]. The loss of Hypocretin which is a neuropeptide found in lateral and posterior hypothalamus reported to be associated with Narcolepsy. Sleepiness in narcolepsy is irresistible and can occur while eating, walking, talking [17]. Oftentimes cataplexy is triggered by strong emotions [17]. Narcolepsy diagnosis is made by clinical history, overnight polysomnography followed by Multiple sleep latency test [17]. Treatment is multimodal with pharmacological and nonpharmacological approaches. Although there is no FDA approved narcolepsy medication for children, Modafinil and Armodafinil are first-line treatment for narcolepsy in patients older than 17 and are used in children off label [17]. Commonly used stimulants are methylphenidate and amphetamines [17].
   2.2 Idiopathic Hypersomnia (IH) manifests chronic excessive daytime sleepiness despite adequate sleep duration and sleep inertia which is difficulty waking from sleep after night-time sleep or a nap [16]. True prevalence is not known due to diagnostic uncertainty [16]. Compared to narcolepsy, sleep paralysis, or hypnogogic hallucinations are less common idiopathic hypersomnia (IH) [16]. Limited data is available about the pathophysiology, diagnosis, and treatment of IH that leads frustration for both physician and patient [16]. Multiple sleep latency test can be used to diagnose IH although

it is imperfect. Recently (in 2021) a mixture of calcium, magnesium, potassium, and sodium oxybates received FDA approval for IH treatment [16].
2.3 Klein–Levin syndrome (KLS) also develops during adolescence and is characterized by periodic episodes of hypersomnia with cognitive dysfunction, behavioral changes, and psychiatric disturbances [16]. Periodic episodes distinguish KLS from other hypersomnia disorders. Hypersomnia occurs during active periods and can be completely resolved in between episodes [16]. KLS is characterized with encephalopathy as well as excessive sleepiness during episodes [16]. Supportive therapy is the mainstay for management of KLS management.
3. Genetic syndromes can be associated with EDS for example Prader Willi, Down syndrome, and myotonic dystrophy.
   3.1 Prader Willi (defects at q11–13 on Chromosome 15) prevalence is reported to be about 1 in 16,000 live births [18]. This genetic syndrome is characterized by unique behavioral profile and associated with developmental delay and intellectual disability. During childhood, hyperphagia leads to obesity and obesity-related problems [18]. Excessive daytime sleepiness reported to be associated with emotional behavioral disturbance in children with Prader Willi [18].
   3.2 Down Syndrome (Trisomy 21) Children with Down syndrome have higher prevalence of OSA and sleep disturbance [19]. Sleep problems reported as high as 74% [19]. Given its high prevalence, children with Down syndrome should be screened for snoring [20]. Polysomnography is useful at diagnosing OSA for this population as well. Treatment should be tailored based on underlying cause.
   3.3 Myotonic Dystrophy (MD): Central nervous system abnormalities occur in MD that leads apathy, memory loss, and mental deterioration, whereas serotonin receptor loss of neurons in the dorsal raphe nucleus and superior central nucleus in addition to hypothalamic hypocretin-orexin system dysfunction can lead to short sleep latencies and sleep-onset REM periods [21] EDS can be debilitation and affect quality of life for both patient and parents in patients with MD [21, 22].
4. Respiratory system related causes: Disordered breathing or pulmonary disorders can manifest as EDS.
   4.1 Obstructive sleep apnea (OSA) prevalence is 3–5% in school-age children [23]. Among those who have OSA prevalence of excessive daytime sleepiness reported as high as 14% [23]. American Academy of Pediatrics (AAP) recommends all children to be screened for snoring and complex cases should be referred to specialist in order to diagnose OSA [24]. Polysomnography is the gold standard test to diagnose OSA and first-line therapy is adenotonsillectomy in children [24].
   4.2 Extreme prematurity can also contribute to EDS due to airway obstruction. Griffiths et al. investigated the association between sleep and pulmonary function among 45 children at age 7–9 who were extremely premature at birth [25]. They concluded that lower FEV1 which reflects greater airflow is correlated with excessive daytime sleepiness [25].

5. Behavioral/Mental health conditions:
   5.1 Internet addiction is an emerging epidemic. Recent studies have shown that 64.1% of young adults have mild to moderate addiction to the internet which is also associated with excessive daytime sleepiness [26].
   5.2 Depression/Anxiety: Mental health is also closely linked to sleep disorders. Depression anxiety and sleep disturbance are frequently associated with each other and sometimes it is hard to tell which one comes first. It is very important to address sleep symptoms for a better outcome in mental health management. A cross-sectional study in 1667 adolescents (mean age: 14.8 ± 1.6) concluded that insomnia symptoms are significantly linked to suicidal ideation [4].
   5.3 ADHD: Children with ADHD has much higher prevalence of sleep disorders compared to control group [27]. A case control study recruited children with ADHD reported sleep phenotypes such as Narcolepsy like phenotype, delayed sleep onset insomnia, obstructive sleep apnea, peripheral limb movements, and sleep epileptiform discharges in the ADHD groups [27]. High prevalence of sleep disorder and short attention span could possibly cause excessive daytime sleepiness in ADHD spectrum.
6. Other:
   6.1 Medication: Daytime excessive sleepiness could be related to medications as well. A careful review of the medication list is indicated when assessing daytime sleepiness.
   6.2 Food allergy: On the other hand, narcolepsy type sleepiness is reported to occur frequently with food allergies and allergic conjunctivitis. Kalb et al. observed 848 children during oral food challenge and 106 (12%) of them reported to have narcolepsy-type sleepiness during observation with a higher incidence in those who have eczema [28]. Children with allergic conjunctivitis and their parents are also found to have poor sleep quality on a prospective case-controlled study [29]. In the study, it is speculated that poor sleep could be related to night-time itchiness or due to circadian release of inflammatory markers such as Histamine which is known to play role in sleep-wake dysregulation [29].
   6.3 EDS could also be acquired and can manifest following traumatic brain injury or brain tumors [30, 31]. Sleep disturbances reported in pediatric age with chronic kidney disease and after kidney transplantation [32, 33]. See Table 13.2 for a summary of differential diagnosis outlined in this section.

Table 13.2 Differential diagnosis for excessive daytime sleepiness

Metabolic derangements
  Hypothyroidism
  Diabetes Mellitus
  Iron deficiency
  Restless leg syndrome
Genetic syndromes
  Prader Willi Syndrome
  Down Syndrome
  Myotonic dystrophy
Sleep disorders
  Central disorder of hypersomnolence (CDH)
    Narcolepsy
    Idiopathic hypersomnia
    Klein-Levin syndrome
  Obstructive sleep apnea
  Obesity hypoventilation
Behavioral/mental health conditions
  Internet addiction
  Depression
  Anxiety
  ADHD
Other
  Traumatic brain injury
  Brain tumor
  Fetal alcohol exposure
  Extreme prematurity
  Kidney transplantation

Clinical Pearls

- Excessive daytime sleepiness in pediatric age is a treatable disorder
- Evaluation starts with complete medical history, then focuses on detailed sleep history, physical examination and specific tests such as polysomnography, actigraphy
- Management of excessive daytime sleepiness is tailored based on the underlying diagnosis and comorbidities

Clinical Vignette

C.S. a 14-year healthy male with a 2-year history of excessive daytime sleepiness. C.S. is a freshman in high school with multiple truancies as its challenging to wake up in the mornings despite getting 9–10 h of sleep. His school teachers report frequent episodes of C.S. falling asleep in class daily and struggling academically. C.S. naps daily after school and on weekends. Parents express concern excessive daytime sleepiness has worsened overtime.

References

1. Young TB. Epidemiology of daytime sleepiness: definitions, symptomatology, and prevalence. J Clin Psychiatry. 2004;65(Suppl 16):12–6.
2. Chervin RD, Scammell TE. Approach to the patient with excessive daytime sleepiness. UpToDate. 2024 [cited 2024 Aug 26]. Available from: https://www.uptodate.com/contents/approach-to-the-patient-with-excessive-daytime-sleepiness?search=excessive%20daytime%20sleepiness%20kids&topicRef=14890&source=see_link
3. Bruni O. Approach to a sleepy child: diagnosis and treatment of excessive daytime sleepiness in children and adolescents. Eur J Paediatr Neurol. 2023;42:97–109.
4. Chan NY, Zhang J, Tsang CC, Li AM, Chan JW, Wing YK, Li SX. The associations of insomnia symptoms and chronotype with daytime sleepiness, mood symptoms and suicide risk in adolescents. Sleep Med. 2020;74:124–31.
5. Komada Y, Ishibashi Y, Hagiwara S, Kobori M, Shimura A. A longitudinal study of subjective daytime sleepiness changes in elementary school children following a temporary school closure due to COVID-19. Children. 2021;8(3):183.
6. Mindell JA, Owens JA. A clinical guide to pediatric sleep: diagnosis and management of sleep problems. Lippincott Williams & Wilkins; 2015.
7. Paurthi S. Evaluation of suspected obstructive sleep apnea in children. UpToDate. 2024 [cited 2024 Sep 2]. Available from: https://www.uptodate.com/contents/evaluation-of-suspected-obstructive-sleep-apnea-in-children?search=daytime%20sleepiness%20and%20body%20habitus&source=search_result&selectedTitle=4%7E150&usage_type=default&display_rank=4
8. Tapia IE, Wise MS. Assessment of sleep disorders in children. UpToDate. 2024 [cited 2024 Aug 28]. Available from: https://www.uptodate.com/contents/assessment-of-sleep-disorders-in-children?source=autocomplete&index=2~6
9. Sun E. Epworth Sleepiness Scale. Sleep Foundation. 2023 [cited 2024 Sep 25]. Available from: https://www.sleepfoundation.org/sleep-studies/epworth-sleepiness-scale
10. Freedman N. Quantifying sleepiness. UpToDate. 2024 [cited 2024 Sep 26]. Available from: https://www.uptodate.com/contents/quantifying-sleepiness?search=stanford%20sleepiness%20scale&source=search_result&selectedTitle=1%7E1&usage_type=default&display_rank=1
11. Zhang D, Zhang Z, Li H, Ding K. Excessive daytime sleepiness in depression and obstructive sleep apnea: more than just an overlapping symptom. Front Psych. 2021;12:710435.
12. Bélanger MÈ, Bernier A, Paquet J, Simard V, Carrier J. Validating actigraphy as a measure of sleep for preschool children. J Clin Sleep Med. 2013;9(7):701–6.
13. Hayashi M, Araki S, Kohyama J, Shimozawa K, Iwakawa Y. Sleep development in children with congenital and acquired hypothyroidism. Brain Dev. 1997;19(1):43–9.
14. Silva RA, Ganen AD, Fernandes VD, Evangelista NM, Figueiredo CC, Pacheco LD, Colares GD. Evaluation of sleep characteristics of children and adolescents with type 1 diabetes mellitus. Rev Paul Pediatr. 2021;40:e2020407.
15. DelRosso LM, Mogavero MP, Bruni O, Ferri R. Restless legs syndrome and restless sleep disorder in children. Sleep Med Clin. 2023;18(2):201–12.
16. Dye TJ. Idiopathic hypersomnia and Kleine–Levin syndrome: primary disorders of Hypersomnolence beyond narcolepsy. In: Seminars in Pediatric neurology, vol. 48. WB Saunders; 2023. p. 101082.
17. Nallu S, Guerrero GY, Lewis-Croswell J, Wittine LM. Review of narcolepsy and other common sleep disorders in children. Adv Pediatr. 2019;66:147–59.
18. Choong CS, Nixon GM, Blackmore AM, Chen W, Jacoby P, Leonard H, Lafferty AR, Ambler G, Kapur N, Bergman PB, Schofield C. Daytime sleepiness and emotional and behavioral disturbances in Prader-Willi syndrome. Eur J Pediatr. 2022;181(6):2491–500.
19. Maris M, Verhulst S, Wojciechowski M, Van de Heyning P, Boudewyns A. Sleep problems and obstructive sleep apnea in children with Down syndrome, an overview. Int J Pediatr Otorhinolaryngol. 2016;82:12–5.

20. Fucà E, Costanzo F. Sleep and behavioral problems in Down syndrome: differences between school age and adolescence. Front Psych. 2023; https://doi.org/10.3389/fpsyt.2023.1193176.
21. Hoxhaj D, Pascazio A, Maestri M, Ricci G, Fabbrini M, Torresi FB, Siciliano G, Bonanni E. Excessive daytime sleepiness in myotonic dystrophy: a narrative review. Front Neurol. 2024;15:1389949.
22. Trucco F, Lizio A, Roma E, di Bari A, Salmin F, Albamonte E, Casiraghi J, Pozzi S, Becchiati S, Antonaci L, Salvalaggio A. Association between reported sleep disorders and Behavioral issues in children with myotonic dystrophy type 1—results from a retrospective analysis in Italy. J Clin Med. 2024;13(18):5459.
23. Yu MW, Au CT, Yuen HM, Chan NY, Chan JW, Wing YK, Li AM, Chan KC. Effects of childhood obstructive sleep apnea with and without daytime sleepiness on behaviors and emotions. Sleep Med. 2024;115:93–9.
24. Oliveira VX, Teng AY. The clinical usefulness of sleep studies in children. Paediatr Respir Rev. 2016;17:53–6.
25. Griffiths V, Blinder H, Hayawi L, Barrowman N, Luu TM, Moraes TJ, Parraga G, Santyr G, Thébaud B, Nuyt AM, Katz SL. Sleep-disordered breathing symptoms and their association with structural and functional pulmonary changes in children born extremely preterm. Eur J Pediatr 2023;182(1):155–163.
26. Monteiro F, Simões M, Relva IC. Internet addiction, sleep habits and family communication: the perspectives of a sample of adolescents. In: Healthcare, vol. 11, No. 24. MDPI; 2023. p. 3194.
27. Miano S, Amato N, Foderaro G, Pezzoli V, Ramelli GP, Toffolet L, Manconi M. Sleep phenotypes in attention deficit hyperactivity disorder. Sleep Med. 2019;60:123–31.
28. Kalb B, Jentsch J, Yürek S, Neumann K, Meixner L, Lau S, Niggemann B, Beyer K. Narcolepsy-like sleepiness: a symptom of immediate-type reactions in food-allergic children. J Allergy Clin Immunol Pract. 2023;11(4):1147–53.
29. Li J, Zhang SY, Fan Z, Liu R, Jin L, Liang L. Impaired sleep quality in children with allergic conjunctivitis and their parents. Eye. 2023;37(8):1558–65.
30. Bogdanov S, Brookes N, Epps A, Naismith SL, Teng A, Lah S. Sleep disturbance in children with moderate or severe traumatic brain injury compared with children with orthopedic injury. J Head Trauma Rehabil 2019;34(2):122–131.
31. Helligsoe AS, Weile KS, Kenborg L, Henriksen LT, Lassen-Ramshad Y, Amidi A, Wu LM, Winther JF, Pickering L, Mathiasen R. Systematic review: sleep disorders based on objective data in children and adolescents treated for a brain tumor. Front Neurosci. 2022;16:808398.
32. Kang KT, Lin MT, Chen YC, Lee CH, Hsu WC, Chang RE. Prevalence of sleep disorders in children with chronic kidney disease: a meta-analysis. Pediatr Nephrol 2022;37(11):2571–2582.
33. Yang M, Chuang SY, Kennedy SE. Sleep disturbances in children and adolescents after kidney transplantation. Pediatr Nephrol. 2024;39(5):1577–85.

Suggested Readings

Approach to a Sleepy Child: Diagnosis and Treatment of Excessive Daytime Sleepiness in Children and Adolescents
Bruni O. Approach to a sleepy child: diagnosis and treatment of excessive daytime sleepiness in children and adolescents. Eur J Paediatr Neurol. 2023;42:97–109. https://doi.org/10.1016/j.ejpn.2022.12.009. Epub 2022 Dec 31. PMID: 36608412.
Sleep Disorders in Children
Ward T, Mason TB 2nd. Sleep disorders in children. Nurs Clin North Am. 2002;37(4):693–706. https://doi.org/10.1016/s0029-6465(02)00032-4. PMID: 12587368.
Review of Narcolepsy and Other Common Sleep Disorders in Children
Nallu S, Guerrero GY, Lewis-Croswell J, Wittine LM. Review of narcolepsy and other common sleep disorders in children. Adv Pediatr Infect Dis. 2019;66:147–59. https://doi.org/10.1016/j.yapd.2019.03.008. Epub 2019 Apr 25. PMID: 31230690.

Part IV

Sleep in Special Populations

Neurological Disorders

14

Daniel Rongo, Rabab Naqvi, Shan Luong, and Sejal V. Jain

Neurological Disorders and Sleep

Sleep and neurological conditions have a bidirectional relationship in their pathophysiology. Sleep dysfunction can exacerbate neurological conditions, while improvements in sleep can lead to better neurological outcomes. Conversely, worsening neurological conditions can negatively impact the quality and duration of sleep. This chapter will focus on the special considerations that should be taken into account when caring for patients with neurological disorders.

Seizures and Epilepsy

Interface of Sleep and Epilepsy

Interictal epileptiform discharges are primarily propagated during non-REM sleep and are less common during REM sleep. Seizures tend to occur most frequently during N2 and N3 stages of sleep and are rare during REM sleep. Conversely, a

Fig. 14.1 Cyclic relationship between sleep quality and seizure control

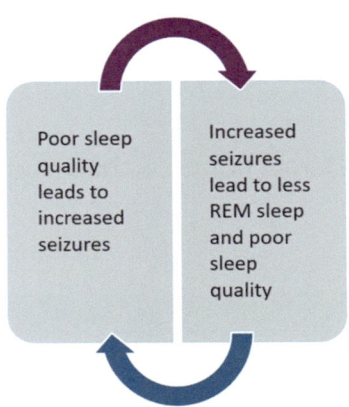

nocturnal seizure can disrupt sleep quality, particularly by reducing REM sleep, which may in turn increase the likelihood of subsequent nocturnal seizures. This creates a vicious cycle, where sleep disruption and seizures exacerbate each other (Fig. 14.1).

Sleep Disorders in Epilepsy

Sleep disorders are common in children with epilepsy, with insomnia being one of the most prevalent. Although there are limited studies directly evaluating the prevalence of insomnia in this population, several questionnaire-based studies have reported poor sleep quality in children with epilepsy. The prevalence of sleep disturbances is noted to be 45% in children with new onset epilepsy, compared to their siblings. This prevalence is even higher in children with cognitive impairment. Most studies suggest that sleep maintenance—the ability to stay asleep—is the most challenging aspect of sleep for these children [1]. In terms of treatment, melatonin is one of the most used hypnotics in children with epilepsy [2]. A prospective, placebo-controlled study by Jain et al. found that sustained-release melatonin significantly reduced sleep onset latency and night awakenings, improving overall sleep quality [3].

OSA is one of the more commonly recognized sleep disorders in epilepsy. According to questionnaire-based studies and mostly retrospective studies in patients presenting to sleep labs for evaluation, the prevalence of OSA in epilepsy ranges from 9% to 65%. The variation in these prevalence rates is primarily due to the differences in the populations studied and the methods used for detection [4]. Children with both epilepsy and OSA tend to experience more frequent and severe sleep disturbances compared to those with OSA alone. Positive Airway Pressure (PAP) treatment has been shown to improve seizure frequency and reduce epileptiform discharges [5]. Similarly, adenotonsillectomy has also been associated with a reduction in seizure frequency [6], further supporting the need to address OSA in these patients.

Hypersomnia is a commonly recognized symptom in children with epilepsy. In one study, 70% of child neurologists attributed excessive sleepiness to antiepileptic medications [2]. However, prospective studies that utilized both subjective and objective tools have shown that sleepiness in these children is often correlated with other conditions, including sleep apnea, restless leg syndrome, habitual snoring, and anxiety [7].

Parasomnias are also commonly seen in children with epilepsy. Certain types of nocturnal seizures can be difficult to distinguish from parasomnias based solely on their descriptions. In such cases, prolonged video EEG monitoring may be necessary to evaluate multiple episodes and identify stereotypes which are suggestive of seizures rather than parasomnia.

About 5–10% of individuals referred to a sleep lab for evaluation are found to have an elevated periodic limb movement index (PLMI) [5]. However, there is limited data regarding the treatment of PLMD in children with epilepsy.

Central sleep apnea and narcolepsy are rarely reported sleep disorders in children with epilepsy.

Sleep and Antiepileptic Therapies

Several studies have evaluated the effects of antiepileptic treatments on sleep [8]. Gabapentinoids, such as gabapentin and pregabalin, have been shown to reduce sleep latency and improve sleep efficiency in individuals with epilepsy. Vagus nerve stimulation (VNS), which is used to treat refractory epilepsy, has been associated with worsening both central and obstructive sleep apnea in some patients. Conversely, successful epilepsy surgery and the ketogenic diet have been found to improve sleep quality due to better seizure control.

Sleep-Related Epilepsy Disorders

Several epilepsy syndromes are characterized by specific patterns of epileptiform discharges and seizures that occur during sleep [7]. Examples include:

1. Self-limited Epilepsy with Centrotemporal Spikes (SeLECTS) (formerly known as Benign Epilepsy with Centrotemporal Spikes (BECTS)): This is a common childhood epilepsy syndrome typically presenting with seizures that often occur during sleep, particularly in the early stages of NREM sleep.
2. Sleep-Related Hypermotor Epilepsies (SHE) (formerly known as nocturnal frontal lobe epilepsy (NFLE)): Characterized by motor seizures that occur predominantly during sleep, these are often hyperkinetic and may involve complex behaviors during the night, such as vocalizations or limb movements.
3. Developmental Epileptic Encephalopathy with Spike-Wave Activation in Sleep (DEE-SWAS) and Epileptic Encephalopathy with Spike Wave Activation in Sleep (EE-SWAS): disorders characterized by cognitive regression with marked activation of interictal epileptiform discharges (IEDs) in NREM sleep.

Headaches

Headache and Sleep Interface

Headaches are one of the most common neurological disorders presenting at the primary care clinics. Several primary headache syndromes are closely associated with specific sleep stage. These disorders include migraines, cluster headaches, and chronic paroxysmal hemicranias. Migraines tend to occur around awakenings or during nocturnal sleep. The attacks are more likely to occur during REM sleep or during arousals from slow wave sleep. In more than half of the subjects, the prodrome of migraine, which is seen 24 h before an attack, includes sleep disturbances. Additionally, occurrence of cluster headaches follows a strong circadian pattern which is same for the same individual. Most of these attacks are believed to occur out of REM sleep, even though occurrence out of other sleep stages have been seen. Surprisingly, the headaches affect sleep structure minimally [9].

Another evidence that primary headaches are associated with intrinsic sleep patterns comes from studies on melatonin. Melatonin levels are reduced in patients with migraine, menstrual migraine, and cluster headaches. Additionally, during a cluster period, melatonin delay is noted in peak melatonin levels. Hence, using melatonin and changing sleep-wake cycle has been a strategy for treatment of cluster headaches [9].

Sleep Disorders in Headaches

Several studies suggest that, in children with headaches, increased sleep problems are noted. In a large sample of school-age children with migraine, tension-type headache and healthy controls, twice as many children with migraine and tension-type headaches had shorter sleep duration and prolonged sleep latency compared to controls. Children with headache had higher difficulty falling asleep as well as fear or anxiety related to sleep. Increased night-time awakenings were also noticed in headache groups. In addition, about a third of the children had restless sleep in the headache groups, sleep breathing problems in the migraine group, and sleep talking, bruxism, and frightening dreams in the migraine group [10]. Similarly, in a school-population-based study, children with migraine had higher sleep problems, increased sleepiness, and preferred evening hours. Poor sleep was reported as the most frequent trigger [11]. Another study showed that snoring, parasomnia, daytime sleepiness, and sweating during sleep were more common in children with migraine as compared to non-migraine headaches [12]. In adolescents with migraine, insufficient total sleep, daytime sleepiness, difficulty falling asleep, and frequent night-time awakenings are seen. In this population, sleep quality is significantly affected by headache pain intensity and duration of headache [13]. Importantly, in a study where a group of children with migraines, who received sleep hygiene instructions, reported shorter headache duration at 6-month follow up, as compared to children without sleep instructions [14].

Despite the compelling information, limited data exist for polysomnographic identified sleep disorders in patients with primary headache disorder. In a retrospective study, polysomnography results from children with migraine, tension-type headache, and non-specific headaches were evaluated. Children with tension-type headache had two-fold increased risk of bruxism. Children with migraine more commonly had obstructive sleep apnea (OSA) as compared to children with tension-type headache. Presence or absence of OSA was not associated with migraine severity. However, patients with severe migraine had shorter total sleep time, shorter REM sleep, reduced slow wave sleep, and increased sleep latency and arousals [15].

When we consider other sleep disorders, four times more children with migraine had restless leg syndrome compared to healthy controls. More than a quarter of the children with migraine had a polysomnographic periodic limb movement index of more than 5 per hour which was associated with a higher frequency, intensity, duration of pain; higher functional impairments and lower response to prophylactic treatment of migraine [14]. Moreover, parasomnias such as night terror, sleep walking, and enuresis are also associated with migraines.

Even though adult studies show that multiple sleep disorders are associated with headaches, limited evidence exists in children. Nevertheless, headaches are reported in up to 70% of adults with OSA and is improved with OSA treatment with continuous positive airway pressure (CPAP). Both tension-type headaches and migraines have significantly high prevalence in narcolepsy [14].

Children with headaches and comorbid sleep issues have a higher risk of developing chronic conditions as they age. Headache and sleep pathologies serve as predictors for likelihood of progression in the future. Headache-free individuals with insomnia had increased risk of developing headaches, further emphasizing the bidirectional relationship of these conditions. [16]

The treatment of headaches focuses on the treatment of both the headache and the comorbid sleep-related issues. Studies show that the use of improved sleep hygiene is known to reduce the duration and frequency of attacks. In those with obstructive sleep apnea, the use of CPAP has shown to significantly improve headaches when using regularly [17].

This data provides a compelling argument that in children with headache, presence of sleep disorders should be evaluated. The identification and treatment of sleep disorder has the potential to improve headaches. Additionally, headaches could also be a presenting feature of a sleep disorder. Hence, accurate diagnosis may be required for adequate management.

Cerebral Palsy

Cerebral palsy (CP) is a neurological disorder caused by prenatal, perinatal, or postnatal brain injury that significantly affects motor function, sensory processing, cognition, behavior, and communication. The relationship between sleep disturbance and CP is complex. Poor sleep in patients with CP can impair their development and daily function. It also impacts the quality of life of the child's caregivers, as poor sleep can exacerbate the patient's symptoms and increase their requirement for assistance and supervision [18].

Sleep Disorders in CP

Sleep disorders represent a major yet often underrecognized issue in this population, affecting an estimated 23–46% of children with CP [18, 19]. Multiple comorbidities in patients with CP can disrupt night-time sleep, including nocturnal seizures, abnormal muscle tone, pain, dysregulated sleep/wake cycles, and the need for specialized positioning equipment (Fig. 14.2). Pain is particularly significant, affecting up to 74% of children with CP. A correlation has been found between the intensity of pain and the severity of sleep disruption [20].

Patients with CP have been found to have a higher prevalence of sleep disorders, including obstructive sleep apnea (OSA) and periodic limb movement disorder (PLMD), which can negatively affect both the quality and quantity of their sleep (Fig. 14.1). In a study of 215 patients with epilepsy and/or CP, the Pediatric Sleep Questionnaire (PSQ) scores were significantly higher in patients with CP (58%) and those with CP and epilepsy (67%) [21]. PLMD also has a higher prevalence in patients with CP. A retrospective review of polysomnography with leg electromyography in patients with neuromuscular dystrophies and CP revealed a higher prevalence of periodic limb movements (PLMs) in these populations compared to other clinic-referred pediatric populations [22].

Additionally, comorbid conditions that affect sleep may also be present. Gastroesophageal reflux (GERD) is very common, affecting up to 90% of patients with CP, and may cause nocturnal awakenings due to acid reflux (Fig. 14.2). This can be exacerbated by laying supine, leading to microarousals and reduced restorative sleep. Furthermore, epilepsy is preset in up to 41% of individuals with CP [18]. These conditions can mimic other conditions, often going unrecognized and presenting as awakening episodes, jerking movements, repetitive limb movements, or vocalizations, potentially resembling other sleep conditions such as parasomnias or periodic limb movement disorder and should be considered as differential diagnosis.

Diagnosis and management approaches should be individualized, taking into account each patient's unique combination of comorbidities and complications. Parents of patients with CP often consider sleep problems as an inherent part of the disorder, which may hamper diagnosis [23]. Hence, primary care providers can utilize screening tools such as the Pediatric Sleep Questionnaire. These questionnaires

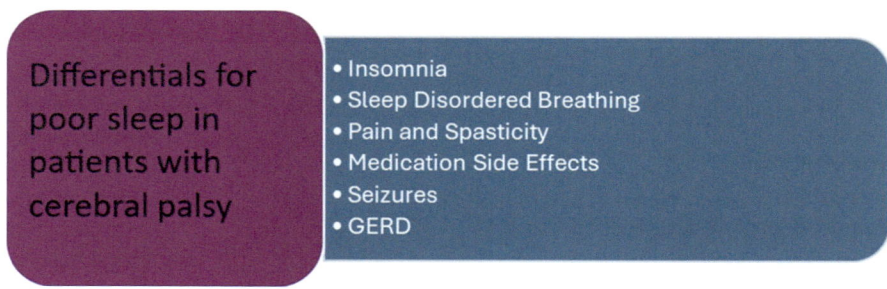

Fig. 14.2 Differential for poor sleep in patients with CP

help quantify sleep quality in a standardized manner. In fact, parental reporting of sleep disturbances is one of the most reliable predictive markers for the diagnosis of OSA. While, positive airway pressure (PAP) therapy remains the gold standard for treating obstructive sleep apnea (OSA), but it has been found to potentially unmask or induce periodic limb movements in sleep (PLMS), which may lead to treatment failure in patients requiring PAP therapy [24]. Additionally, in patients with CP and spastic quadriplegia, intrathecal baclofen therapy has shown promise in improving both pain control and sleep-disordered breathing by better managing muscle tone. Meanwhile, there is limited evidence-based guidance regarding pharmacologic therapy for insomnia in this population. Medications such as clonidine and gabapentin have shown efficacy in treating sleep disturbances in neurodevelopmental disorders. However, careful consideration must be given to potential side effects, including increased daytime sedation, worsening of sleep-disordered breathing, behavioral changes, and potential drug interactions. Further research is needed to assess the implications of PLMS in this population.

Autism Spectrum Disorder

Autism spectrum disorder (ASD) and sleep disorders share important overlaps that should be considered in the primary care setting. Sleep problems are extremely common in children with ASD, affecting 50–80% of patients. Patients with autism have been found to have a 34–58% prevalence of sleep apnea as compared to the general population (1–4%) [25]. These sleep issues include difficulty falling asleep (sleep onset insomnia), frequent night-time awakenings (sleep maintenance insomnia), and irregular sleep-wake schedules [26, 27]. These sleep disturbances tend to persist into adolescence and adulthood, potentially worsening daytime functioning, often manifesting as increased aggression, self-injury, heightened anxiety, hyperactivity, and inattention. These issues can also significantly impact the quality of life and the quality of sleep for their close family members, including parents and siblings [28].

Screening for sleep-related conditions in ASD should be an essential part of primary care visits. Key components of sleep evaluation include sleep environment, bedtime routines, and nighttime awakenings. Validated sleep assessment tools can aid in the evaluation of sleep, such as Modified Simonds and Parraga Sleep Questionnaire, Children's Sleep Habits Questionnaire, Pediatric Sleep Clinical Global Impressions Scale (CGI-S) [29, 30]. Additionally, it is important to screen for sleep apnea, particularly in children with tonsillar hypertrophy and obesity.

The management of sleep-related issues in patients with ASD is multifaceted. The first step is behavioral interventions, such as educating the parents to establish regular bedtime routines with the aid of visual schedules (Fig. 14.3) and providing positive reinforcement. In cases where behavioral interventions alone are insufficient, pharmacologic treatment can be considered. Melatonin is one of the most commonly used pharmacologic agents for children with ASD who struggle with sleep initiation and circadian rhythm disruption. Clonidine has shown effectiveness

Fig. 14.3 Example of a visual schedule

in reducing sleep latency, night-time awakenings, and aggression. [31] Other pharmacological agents, such as antihistamines (e.g., diphenhydramine), melatonin receptor agonists (e.g., ramelteon), and atypical antipsychotics (e.g., risperidone, aripiprazole) should be targeted toward behavior-specific issues as indicated. Additionally, a collaborative, multidisciplinary approach involving the primary care pediatrician, sleep specialist, and behavioral analyst is important for managing sleep issues in children with ASD [32].

Special Considerations for Diagnosis and Treatment

Recognizing and addressing the underlying comorbid sleep-related condition in children with neurological disorders is crucial for improving both their neurologic outcomes and overall quality of life. Given the bidirectional relationship between sleep disturbances and many neurological conditions, early detection and intervention in the primary care setting can significantly enhance treatment success.

Evaluation and diagnosis of sleep disorders in children with neurological disorders are similar to those in typically developing children without neurological disorders. Screening tools can improve the detection of sleep issues in these patients. Studies show that it is not possible to diagnose a specific sleep disorder based solely on presenting complaints. Therefore, the initial approach to a complaint of difficulty

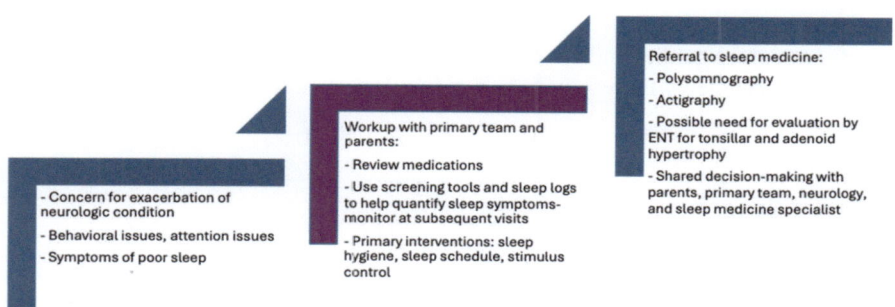

Fig. 14.4 Management for patients with chronic neurologic conditions with concern for sleep pathology

falling asleep, nocturnal awakenings, or sleepiness should involve evaluating sleep hygiene and the sleep-wake schedule (Fig. 14.4). If these are addressed and symptoms are ongoing, or if symptoms specific to a sleep disorder, such as snoring, witnessed apnea, or restless leg syndrome, are noted, a referral to a sleep-disorder specialist should be made.

Here are some of the screening tools which may be used in these populations to help identify the seep problem. One such tool is a quick 5-item BEARS sleep screening tool developed for children ages 2–12 [33]. It stands for bedtime issues, excessive daytime somnolence, night awakenings, regularity and duration of sleep, and snoring. Though scoring is largely dependent on the clinician's interpretation, it can help prompt a referral to a Sleep specialist to further evaluate sleep disorders. For our patient in the above case, he would have scored positive for at least 2 out of the 5 items suggestive of a sleep disorder. Another screener is the PSQ, or Pediatric Sleep Questionnaire. It has shown to have a sensitivity of 0.81–0.85 and a specificity of 0.87 in detection of sleep-related breathing disorders [25]. The patients with positive questionnaire may be referred further evaluation by a sleep specialist or polysomnography. Detail evaluations and management of sleep disorders are included in other chapters.

Pearls/Take-Home Points

- Sleep disorder and neurological disorders, such as cerebral palsy, autism spectrum disorder, epilepsy, and headaches, share a complex, bi-directional relationship requiring management of both aspects for effective treatment.
- Patients with neurologic conditions should be evaluated for comorbid sleep-related conditions (Fig. 14.4).
- Early recognition of sleep disorders in these neurological conditions is crucial, as they can have detrimental long-term effects on health.
- A multidisciplinary approach, involving collaboration between sleep specialists, neurologists, and primary care physicians is essential in managing patients with these comorbid conditions (Fig. 14.4).

- Parents of children with neurologic conditions should be educated on the importance of sleep quality and quantity to prevent further complications. They should also be provided with resources to monitor, report, and manage poor sleep.

Patient Case

A 7-year-old boy with a history of cerebral palsy presented to his primary care provider for an annual check-up. He had mild intellectual disability and epilepsy, for which he was taking levetiracetam. His mother reported that he did not seem to sleep well at night. She had heard loud snoring and, on occasion, had noticed pauses in his breathing during sleep. His teachers also informed her that he had been taking naps during class since the beginning of the academic year. Additionally, she observed that he fell asleep easily in the car on the drive home from school in the afternoons.

His mother noted that since starting levetiracetam, his difficulty falling asleep improved. After his evening dose of levetiracetam, he usually fell asleep within 30 min, whereas previously it could take him 1–2 h to fall asleep.

On physical examination, the head was normocephalic. The patient exhibited slight irritability and poor eye contact. There was decreased muscle tone and strength on the right side of the body. The nasal mucosa was mildly erythematous with clear discharge. Tonsils were moderately enlarged. Cervical lymphadenopathy was not noted. Lungs were clear to auscultation.

Differential Diagnosis

This child presents with persistent daytime sleepiness in the setting of epilepsy and cerebral palsy. His differential diagnosis includes:

- Sleep-related breathing disorders (obstructive vs. central sleep apnea)
- Refractory seizures/nocturnal seizures
- Medication side-effects
- Insufficient sleep/insomnia
- Other medical conditions (e.g., hypothyroidism, anemia)

Plan

He was referred for, and completed, an overnight polysomnogram, which revealed moderate obstructive sleep apnea with an AHI of 6 events per hour. He was subsequently referred to ENT and underwent flexible nasopharyngoscopy, which showed bilateral grade 3 adenoids occupying approximately 60% of the nasopharyngeal airway. He then underwent a bilateral adenotonsillectomy (T&A). A repeat polysomnogram performed 3 months later showed an improved AHI of 1 event per hour, with resolution of daytime sleepiness.

Discussion

Pediatric patients with neurological disorders are often at an increased risk for sleep disorders. Sleep is crucial for both physical and cognitive development, including emotional regulation and learning [34]. Disturbances in sleep can have lasting effects, such as disruption of normal brain development, impaired physical growth, and reduced clearance of neurotoxic proteins [35, 36]. When compounded by the effects of neurologic disorders (e.g., epilepsy, cerebral palsy, autism, headaches), sleep disorders can further exacerbate the already detrimental impact on development, quality of life, mental health, and symptomatology in these patients [37, 38]. So these disorders should be actively evaluated, diagnosed, and treated. This may not only lead to improvement in sleep but also improve symptoms of the neurological disorder and quality of sleep.

Disclosures SVJ is one of the owners and the treasurer of Scientific Research Group, PLLC.

References

1. Jain SV. Epilepsy and sleep. In: Pediatric epilepsy, diagnosis and treatment. Demos Medical Publishing; 2017.
2. Jain SV, Simakajornboon N, Glauser TA. Provider practices impact adequate diagnosis of sleep disorders in children with epilepsy. J Child Neurol. 2012; https://doi.org/10.1177/0883073812449692.
3. Jain SV, Horn PS, Simakajornboon N, et al. Melatonin improves sleep in children with epilepsy: a randomized, double-blind, crossover study. Sleep Med. 2015;16(5):637–44. https://doi.org/10.1016/j.sleep.2015.01.005.
4. Winsor AA, Richards C, Bissell S, Seri S, Liew A, Bagshaw AP. Sleep disruption in children and adolescents with epilepsy: a systematic review and meta-analysis. Sleep Med Rev. 2021;57:101416. https://doi.org/10.1016/j.smrv.2021.101416.
5. Kaleyias J, Cruz M, Goraya JS, et al. Spectrum of polysomnographic abnormalities in children with epilepsy. Pediatr Neurol. 2008;39(3):170–6. https://doi.org/10.1016/j.pediatrneurol.2008.06.002.
6. Segal E, Vendrame M, Gregas M, Loddenkemper T, Kothare SV. Effect of treatment of obstructive sleep apnea on seizure outcomes in children with epilepsy. Pediatr Neurol. 2012;46(6):359–62. https://doi.org/10.1016/j.pediatrneurol.2012.03.005.
7. Liu WK, Kothare S, Jain S. Sleep and epilepsy. Semin Pediatr Neurol. 2023;48:101087. https://doi.org/10.1016/j.spen.2023.101087.
8. Jain SV, Glauser TA. Effects of epilepsy treatments on sleep architecture and daytime sleepiness: an evidence-based review of objective sleep metrics. Epilepsia. 2014;55(1):26–37. https://doi.org/10.1111/epi.12478.
9. Dodick DW, Eross EJ, Parish JM, Silber M. Clinical, anatomical, and physiologic relationship between sleep and headache. Headache. 2003;43(3):282–92. https://doi.org/10.1046/j.1526-4610.2003.03055.x.

10. Bellini B, Panunzi S, Bruni O, Guidetti V. Headache and sleep in children. Curr Pain Headache Rep. 2013;17(6):335. https://doi.org/10.1007/s11916-013-0335-x.
11. Bruni O, Russo PM, Ferri R, Novelli L, Galli F, Guidetti V. Relationships between headache and sleep in a non-clinical population of children and adolescents. Sleep Med. 2008;9(5):542–8. https://doi.org/10.1016/j.sleep.2007.08.010.
12. Isik U, Ersu RH, Ay P, et al. Prevalence of headache and its association with sleep disorders in children. Pediatr Neurol. 2007;36(3):146–51. https://doi.org/10.1016/j.pediatrneurol.2006.11.006.
13. Gilman DK, Palermo TM, Kabbouche MA, Hershey AD, Powers SW. Primary headache and sleep disturbances in adolescents. Headache. 2007;47(8):1189–94. https://doi.org/10.1111/j.1526-4610.2007.00885.x.
14. Dosi C, Figura M, Ferri R, Bruni O. Sleep and headache. Semin Pediatr Neurol. 2015;22(2):105–12. https://doi.org/10.1016/j.spen.2015.04.005.
15. Vendrame M, Kaleyias J, Valencia I, Legido A, Kothare SV. Polysomnographic findings in children with headaches. Pediatr Neurol. 2008;39(1):6–11. https://doi.org/10.1016/j.pediatrneurol.2008.03.007.
16. Tiseo C, Vacca A, Felbush A, et al. Migraine and sleep disorders: a systematic review. J Headache Pain. 2020;21(1):126. https://doi.org/10.1186/s10194-020-01192-5.
17. Guidetti V, Dosi C, Bruni O. The relationship between sleep and headache in children: implications for treatment. Cephalalgia. 2014;34(10):767–76. https://doi.org/10.1177/0333102414541817.
18. Dutt R, Roduta-Roberts M, Brown CA. Sleep and children with cerebral palsy: a review of current evidence and environmental non-pharmacological interventions. Children (Basel). 2015;2(1):78–88. https://doi.org/10.3390/children2010078.
19. Simard-Tremblay E, Constantin E, Gruber R, Brouillette RT, Shevell M. Sleep in children with cerebral palsy: a review. J Child Neurol. 2011;26(10):1303–10. https://doi.org/10.1177/0883073811408902.
20. Ostojic K, Paget SP, Morrow AM. Management of pain in children and adolescents with cerebral palsy: a systematic review. Dev Med Child Neurol. 2019;61(3):315–21. https://doi.org/10.1111/dmcn.14088.
21. Rosenbaum P, Stewart D. The World Health Organization International Classification of Functioning, Disability, and Health (ICF): a model to guide clinical thinking, practice and research in the field of cerebral palsy. Dev Med Child Neurol. 2004;46(1):3–9. https://doi.org/10.1111/dmcn.13091.
22. Nisbet LC, Davey MJ, Nixon GM. Periodic limb movements during sleep in children with neuromuscular disease or cerebral palsy—an important potential contributor to sleep-related morbidity. Sleep Med. 2024;121:58–62. https://doi.org/10.1016/j.sleep.2024.06.017.
23. Morley A. Cerebral palsy and sleep disordered breathing. Breathe (Sheff). 2016;12(4):357–63. https://doi.org/10.1183/20734735.012016.
24. Pai V, Khatwa U, Ramgopal S, Singh K, Fitzgerald R, Kothare SV. Prevalence of pediatric periodic leg movements of sleep after initiation of PAP therapy. Pediatr Pulmonol. 2014;49(3):252–6. https://doi.org/10.1002/ppul.22802.
25. Zaffanello M, Piacentini G, Nosetti L, et al. Sleep disordered breathing in children with autism spectrum disorder: an in-depth review of correlations and complexities. Children (Basel). 2023;10(10):1609. https://doi.org/10.3390/children10101609.
26. Cortese S, Faraone SV, Konofal E, Lecendreux M. Sleep in children with attention-deficit/hyperactivity disorder: meta-analysis of subjective and objective studies. J Am Acad Child Adolesc Psychiatry. 2009;48(9):894–908. https://doi.org/10.1097/CHI.0b013e3181ac09c9.
27. Soulders MC, Mason TBA, Valladares O, et al. Sleep behaviors and sleep quality in children with autism spectrum disorders. Sleep. 2009;32(12):1566–78.
28. McLean KJ, Eack SM, Bishop L. The impact of sleep quality on quality of life for autistic adults. Res Autism Spectr Disord. 2021;88:101849. https://doi.org/10.1016/j.rasd.2021.101849.
29. Malow BA, Connolly HV, Weiss SK, et al. The Pediatric Sleep Clinical Global Impressions Scale—a new tool to measure pediatric insomnia in autism spectrum disorders. J Dev Behav Pediatr. 2016;37(5):370–6. https://doi.org/10.1097/DBP.0000000000000307.

30. Johnson CR, Turner KS, Foldes EL, Malow BA, Wiggs L. Comparison of sleep questionnaires in the assessment of sleep disturbances in children with autism spectrum disorders. Sleep Med. 2012;13(7):795–801. https://doi.org/10.1016/j.sleep.2012.03.005.
31. Ming X, Gordon E, Kang N, Wagner GC. Use of clonidine in children with autism spectrum disorders. Brain Dev. 2008;30(7):454–60.
32. Johnson KP, Zarrinnegar P. Autism spectrum disorder and sleep. Psychiatr Clin North Am. 2024;47(2):199–212. https://doi.org/10.1016/j.psc.2023.06.013.
33. Owens JA, Dalzell V. Use of the 'BEARS' sleep screening tool in a pediatric residents' continuity clinic: a pilot study. Sleep Med. 2005;6(1):63–9. https://doi.org/10.1016/j.sleep.2004.07.015.
34. Gruber R, Cassoff J, Frenette S, Wiebe S, Carrier J. Impact of sleep extension and restriction on children's emotional lability and impulsivity. Pediatrics. 2012;130(5):e1155–61. https://doi.org/10.1542/peds.2012-0564.
35. Jenni OG, O'Connor BB. Children's sleep: an interplay between culture and biology. Pediatrics. 2005;115(1 Suppl):204–16. https://doi.org/10.1542/peds.2004-0815B.
36. Xie L, Kang H, Xu Q, et al. Sleep drives metabolite clearance from the adult brain. Science. 2013;342(6156):373–7. https://doi.org/10.1126/science.1241224.
37. Meltzer LJ. Future directions in sleep and developmental psychopathology. J Clin Child Adolesc Psychol. 2017;46(2):295–301. https://doi.org/10.1080/15374416.2016.1236727.
38. Owens JA. A clinical overview of sleep and attention-deficit/hyperactivity disorder in children and adolescents. J Can Acad Child Adolesc Psychiatry. 2009;18(2):92–102.

The Restless Mind: Sleep Disorders and Mental Health in Children

15

Imran S. Khawaja and Joshua Robbins

Examination/Mental Status Exam

On examination, the patient presented as visibly anxious and restless, with a diminished attention span reflecting his struggle to focus and engage in conversation. His body mass index (BMI) was within the normal range, though physical examination initially indicated nasal obstruction, attributed to enlarged tonsils and adenoids, contributing to his night-time breathing difficulties. The pediatric Epworth Sleepiness Scale revealed a concerning score of 14/24, indicating significant excessive daytime sleepiness [5].

Diagnostic Workup

1. *Polysomnography*: A comprehensive sleep study confirmed moderate obstructive sleep apnea, with an apnea-hypopnea index (AHI) of 12, linked to his enlarged tonsils. This finding led to a tonsillectomy and adenoidectomy, which successfully resolved the OSA symptoms.
2. *Multiple Sleep Latency Test (MSLT)*: The MSLT demonstrated excessive daytime sleepiness, with a mean sleep latency of 6 min, underscoring the profound impact of his initial sleep disturbances on daytime alertness.
3. *Restless Legs Syndrome Assessment*: Evaluation indicated a worsening of RLS symptoms following the initiation of antidepressant therapy, highlighting the connection between the medication and exacerbated sleep disturbances [6].

I. S. Khawaja (✉)
Department of Psychiatry, University of Oklahoma, Norman, OK, USA

MD TruCare, Grapevine, TX, USA

J. Robbins
Franciscan University of Steubenville, Steubenville, OH, USA

Discussion

Sleep Deprivation and Mental Health

Chronic sleep deprivation significantly affects children, with disrupted sleep patterns leading to impaired cognitive functions and emotional regulation. Sleep-deprived children often exhibit increased impulsivity and hyperactivity, complicating existing ADHD symptoms [1]. Additionally, sleep deprivation elevates irritability and anxiety, underscoring the intricate link between sleep quality and mental health [8].

Delayed Sleep Phase Syndrome (DSPS)

Delayed Sleep Phase Syndrome poses a substantial challenge for many children, characterized by a persistent misalignment between the child's biological clock and socially dictated schedules. This misalignment often exacerbates anxiety and depression as affected children struggle with late sleep onset and difficulty waking up in time for school and social activities [5]. The resulting cycle of sleep deprivation further deteriorates mood and cognitive performance, impacting overall mental health.

Snoring and Obstructive Sleep Apnea (OSA)

OSA, marked by repeated episodes of airway obstruction during sleep, leads to fragmented sleep and hypoxemia. Frequent sleep interruptions are associated with increased risks of mood disorders, including anxiety and depression, as disrupted sleep compromises restorative functions essential for optimal cognitive processing [2].

Restless Legs Syndrome (RLS) and Antidepressants

RLS presents unique challenges, especially when symptoms are aggravated by pharmacologic treatments such as antidepressants. The uncomfortable sensations caused by RLS disrupt sleep patterns, creating a cycle of anxiety and frustration for affected children [4]. Effective management of RLS requires a holistic approach, incorporating careful medication adjustments and behavioral interventions to improve sleep quality [7].

Treatment

1. *Behavioral Interventions*: Tailored Cognitive Behavioral Therapy for Insomnia (CBT-I) and comprehensive sleep hygiene education were implemented. This approach emphasized relaxation techniques and an established bedtime routine to support healthy sleep practices [1].
2. *Medical Management*: Nasal corticosteroids and antihistamines were administered to address nasal obstruction. Adjustments to the patient's antidepressant regimen were also made to mitigate the exacerbation of RLS symptoms [6].
3. *Continuous Positive Airway Pressure (CPAP)*: Although initially considered, CPAP was rendered unnecessary following the successful tonsillectomy and adenoidectomy, which resolved the moderate OSA.
4. *Chronotherapy*: Gradual adjustments in sleep timing were employed to help realign the child's circadian rhythm and manage his delayed sleep phase, improving his alignment with daily routines [8].

Results and Follow-Up

Three months post-treatment, the patient showed notable improvements in attention span and emotional stability, enhancing his overall quality of life. Follow-up polysomnography showed a decrease in AHI to 3, confirming the resolution of OSA symptoms. The revised antidepressant regimen effectively managed the patient's RLS symptoms, improving both sleep quality and daytime functionality.

Sleep and Mental Health in Children

Given the clear connection between sleep and mental health in children, addressing sleep issues can be a critical component of mental health treatment. Interventions such as cognitive-behavioral therapy for insomnia (CBT-I), chronotherapy, and sleep hygiene education have shown promise in improving sleep quality and mental health outcomes. The research underscores the importance of a multidisciplinary approach to addressing sleep disorders in children with mental health issues, combining behavioral interventions with pharmacological support when necessary.

Interventions and Therapeutic Approaches

The biological pathways linking sleep and mental health are complex and include the dysregulation of stress hormones, such as cortisol, and neurotransmitter systems, including serotonin and dopamine. Sleep disruptions can lead to increased levels of cortisol, which is associated with stress and anxiety. Moreover, inadequate sleep interferes with the production of serotonin, a neurotransmitter crucial for mood regulation, contributing to symptoms of depression and anxiety.

Biological Mechanisms Linking Sleep and Mental Health

3. Depression and Sleep Patterns: Disturbed sleep is both a symptom and a predictor of depression in children and adolescents. Poor sleep quality can lead to diminished motivation, lowered self-esteem, and social withdrawal, which are common symptoms of depression. Longitudinal studies have found that chronic sleep problems in early childhood may increase the risk of developing depression later in life, underscoring the importance of early intervention.

2. **Anxiety Disorders**: Anxiety is both a cause and a consequence of sleep disturbances in children. Children with anxiety disorders are prone to insomnia, nightmares, and frequent nocturnal awakenings, which can perpetuate a cycle of sleep deprivation and heightened anxiety. Found that cognitive-behavioral interventions targeting sleep in anxious children can improve both sleep quality and anxiety symptoms, suggesting that sleep interventions may have therapeutic benefits for broader mental health issues.

1. ADHD and Sleep: Children with ADHD often experience sleep issues such as delayed sleep onset, shorter sleep duration, and increased sleep fragmentation. These disturbances can exacerbate core ADHD symptoms, including impulsivity, inattention, and hyperactivity [3]. Studies suggest that addressing sleep issues in children with ADHD can lead to improvements in both attention and behavior, highlighting the bidirectional relationship between sleep quality and ADHD symptomatology.

Impact of Specific Disorders on Mental Health

Research indicates that children with chronic sleep deprivation exhibit greater emotional instability, including heightened irritability, aggression, and mood swings. Sleep loss negatively affects the brain regions responsible for emotion regulation, particularly the amygdala and prefrontal cortex, reducing the child's ability to process emotional cues accurately. Furthermore, a lack of sleep amplifies stress responses, which can exacerbate symptoms of anxiety and depression in children.

Sleep Deprivation and Emotional Dysregulation

Sleep plays a crucial role in children's development and maintenance of emotional regulation and cognitive function. Disrupted sleep can have profound effects on mental health, particularly in children who are already vulnerable due to conditions like anxiety, ADHD, and depression.

Clinical History/Case

The patient was an 8-year-old boy with a dual diagnosis of ADHD and generalized anxiety disorder (GAD). He presented to our clinic with his parents due to initial insomnia, which had progressively worsened over 6 months. His sleep difficulties persisted despite behavioral modifications, including a structured bedtime routine and a consistent sleep schedule. The patient reported experiencing a sensation in his legs described as "the creepy crawlies" that caused severe restlessness, compelling him to pace around his room. These symptoms began after initiating an antidepressant prescribed to address his anxiety, suggesting the development of antidepressant-induced Restless Legs Syndrome (RLS) [4].

The patient's parents also reported loud snoring and intermittent gasping during sleep. Physical examination and imaging indicated that these symptoms were due to enlarged tonsils. The patient subsequently underwent a tonsillectomy and adenoidectomy, which successfully resolved the snoring and breathing interruptions, thus alleviating symptoms associated with obstructive sleep apnea (OSA) [2].

The patient's poor sleep quality led to worsening ADHD symptoms, including hyperactivity, inattentiveness, and irritability, which manifested in behavioral issues such as emotional outbursts, reduced classroom participation, and conflicts with peers, teachers, and family. Following the tonsillectomy and adenoidectomy, the patient's sleep quality improved, and his ADHD symptoms gradually lessened, positively impacting his daily functioning.

References

1. Bartel KA, Gradisar M. The relationship between sleep, anxiety, and behavior in children: a review of the literature. Sleep Med Rev. 2019;47:101215. https://doi.org/10.1016/j.smrv.2019.02.006.
2. Cheung JC, Chow CB. Obstructive sleep apnea and its impact on the mental health of children: a systematic review. J Sleep Res. 2020;29(3):e13014. https://doi.org/10.1111/jsr.13014.
3. Cortese S, et al. Sleep problems in children with ADHD: a systematic review and meta-analysis. Sleep Med Rev. 2018;42:27–38. https://doi.org/10.1016/j.smrv.2017.09.001.
4. Guilleminault C, Philip P. Restless legs syndrome and sleep disturbance in children and adolescents. Sleep Med Clin. 2015;10(4):473–84. https://doi.org/10.1016/j.jsmc.2015.07.006.
5. Meltzer LJ, Mindell JA. Sleep and sleep disorders in children and adolescents. Pediatrics. 2006;117(3):e703–12. https://doi.org/10.1542/peds.2005-1686.
6. Owens JA. Insufficient sleep in adolescents and young adults: an opportunity for action. The National Academies Press; 2014. https://doi.org/10.17226/21629.
7. Smith MT, Perlis ML. Sleepodynamics: the interplay between sleep, affect, and performance. J Exp Psychol Gen. 2006;135(3):421–34. https://doi.org/10.1037/0096-3445.135.3.421.
8. Tzischinsky O, Shragai R. The sleep-wake cycle in children and adolescents: implications for sleep disorders. J Fam Issues. 2016;37(4):494–511. https://doi.org/10.1177/0192513X1452872.
9. Verhulst SJ, et al. The relationship between sleep and emotional problems in children and adolescents: a systematic review. Sleep Med Rev. 2015;26:46–55. https://doi.org/10.1016/j.smrv.2014.06.003.

Sleep in Special Populations: Genetic Syndromes

16

Likhita Shaik, Janey Dudley, Daniel Rongo, Elisa Basora, and Anna Wani

Introduction

Sleep disturbances are highly prevalent (nearly 80%) among children with genetic syndromes, and these issues can significantly intensify the developmental, cognitive, and behavioral challenges characteristic to each condition [1, 2]. The type and severity of sleep issues vary depending on the underlying pathophysiologic process involved in causing the sleep disorder. Some children suffer from obstructive sleep apnea (OSA), insomnia, or circadian rhythm disorders. Understanding how genetic abnormalities interact with sleep regulation is essential for clinicians to manage these disorders effectively [3].

This chapter explores various genetic syndromes, the pathophysiological sleep mechanisms causing the sleep disorder and reliable management strategies. More common syndromes, such as Down syndrome, Prader-Willi syndrome, Angelman syndrome, Fragile X syndrome, will be discussed in more detail than others.

L. Shaik · J. Dudley · D. Rongo
Department of Sleep Medicine, UT Southwestern, Dallas, TX, USA
e-mail: likhita.shaik@utsouthwestern.edu; janey.dudley@utsouthwestern.edu; daniel.rongo@utsouthwestern.edu

E. Basora
Department of Sleep Medicine, Children's Health, Dallas, TX, USA
e-mail: elisa.basorarovira@utsouthwestern.edu

A. Wani (✉)
Family and Community Medicine, and Pediatrics, The University of Texas Southwestern Medical Center, Plano, TX, USA

Department of Sleep Medicine, Children's Health, Dallas, TX, USA
e-mail: anna.wani@utsouthwestern.edu

Syndromes with Craniofacial and Airway Abnormalities (Fig. 16.1) [3]

Down Syndrome (Trisomy 21)

Children with Down syndrome (DS) have a significantly increased risk of obstructive sleep apnea (OSA), with prevalence rates ranging from 50% to 80% [3, 4]. Anatomical factors such as macroglossia, midfacial hypoplasia, and adenotonsillar hypertrophy contribute to upper airway obstruction during sleep [5]. A study by Shott et al. reported five times more likelihood of children with DS to develop OSA compared to the general pediatric population [4]. If left untreated, OSA can lead to significant cognitive, behavioral, and cardiovascular complications in patients with DS [5]. Children with DS also exhibit abnormal sleep architecture, characterized by frequent arousals, reduced REM sleep, and shorter durations of slow-wave sleep. These disruptions deleteriously impact daytime functioning, contributing to increased daytime sleepiness, learning difficulties, and behavioral issues [6].

Management: Adenotonsillectomy is often the first-line treatment for OSA in younger children with DS. However, about 30–50% of children experience residual OSA following surgery [7]. In these cases, continuous positive airway pressure (CPAP) therapy is recommended. Other interventions, such as weight management

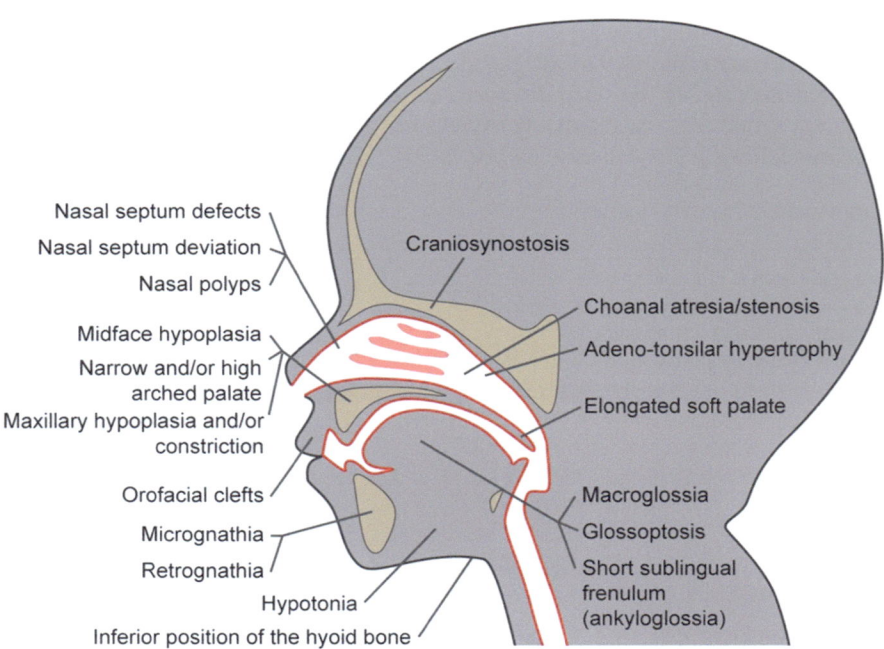

Fig. 16.1 Anatomical features of various Craniofacial and Airway Abnormalities

and orthodontic approaches to treat midfacial hypoplasia, may also improve outcomes [8].

Pierre Robin Sequence (PRS)

Sleep apnea is a common issue associated with PRS, often linked to anatomical challenges like a small lower jaw or cleft palate. Early identification through sleep studies is crucial (Fig. 16.2) [9]. For mild cases, prone sleeping positions can improve airway patency. Severe cases may require interventions such as mandibular distraction (to extend the jaw) or tongue-lip adhesion surgery. In some instances, CPAP therapy is effective in addressing obstructive sleep apnea [9, 10].

Management During Infancy: In infancy, airway monitoring is essential. Prone positioning during sleep helps prevent the tongue from obstructing the airway. Around 30% of babies with PRS may require surgery during their first year to stabilize their breathing. When structural adjustments aren't sufficient, a tracheostomy might be necessary for ensuring adequate airflow [10, 11].

Long-Term Considerations: As children grow, jaw growth can mitigate symptoms naturally for some. If growth isn't sufficient by early childhood or adolescence, further surgical interventions may be necessary. Monitoring and treatment of sleep apnea remain priorities throughout development [11].

Achondroplasia

Achondroplasia, a genetic disorder caused by mutations in the *FGFR3* gene, is the most common form of disproportionate short stature. It is characterized by a

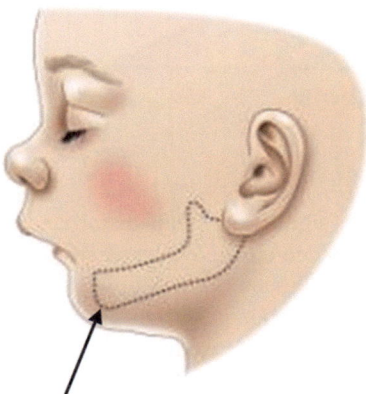
Micrognathia - a small jaw with a receding chin

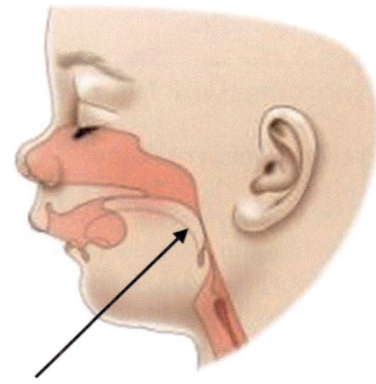
Tongue that is large compared to the jaw, resulting in airway obstruction

Fig. 16.2 PRS airway features

narrow foramen magnum, midface hypoplasia, and upper airway obstruction, predisposing individuals to sleep-disordered breathing (SDB), including obstructive sleep apnea (OSA). Studies estimate OSA prevalence in children with achondroplasia to range from 50% to 75% [12, 13]. The craniofacial abnormalities in achondroplasia lead to reduced airway space, and hypotonia may further exacerbate airway obstruction during sleep. Additionally, brainstem compression due to a narrow foramen magnum can contribute to central sleep apnea (CSA) [14]. Untreated SDB can result in developmental delay, behavioral disturbances, and cardiovascular complications.

Management: Management strategies include early detection through sleep studies and targeted interventions. Mild cases of OSA can be managed with non-invasive methods like continuous positive airway pressure (CPAP). Surgical interventions, such as adenotonsillectomy or midface advancement, are often indicated for more severe cases of OSA. For central sleep apnea caused by foramen magnum stenosis, neurosurgical decompression may be necessary [15, 16]. Long-term monitoring is essential to optimize treatment outcomes and prevent complications.

Apert Syndrome and Other Craniofacial Syndromes

Like DS, children with craniosynostosis syndromes, such as Apert syndrome, are also prone to OSA due to midfacial hypoplasia and restricted airway space. These children may also exhibit disrupted sleep architecture, including reduced REM sleep and frequent night-time arousals [17]. Similar challenges are seen in other craniofacial syndromes like Crouzon syndrome and Pfeiffer syndrome, where midfacial hypoplasia and airway abnormalities contribute to SDB(Fig. 16.3) [18]. Persistent airway obstruction often requires surgical intervention, including adenotonsillectomy and, in severe cases, midface advancement surgery to improve airflow [19].

Syndromes with Neuromuscular Hypotonia

Prader-Willi Syndrome

Prader-Willi syndrome (PWS) is associated with significant sleep disturbances, including sleep disordered breathing (SDB) (prevalence up to 98% excessive daytime sleepiness (EDS) (approximately 30–40%) [20]. Hypotonia, a defining feature of PWS, contributes to the high incidence of OSA by impairing the muscle tone necessary for maintaining airway patency during sleep [21]. Additionally, central sleep apnea (CSA) occurring from hypothalamic dysfunction is also a characteristic of the syndrome [22]. Abnormal circadian rhythms due to delayed melatonin secretion causing disrupted sleep-wake patterns are also common in PWS. Even when nocturnal sleep appears adequate, individuals with PWS often experience EDS [21].

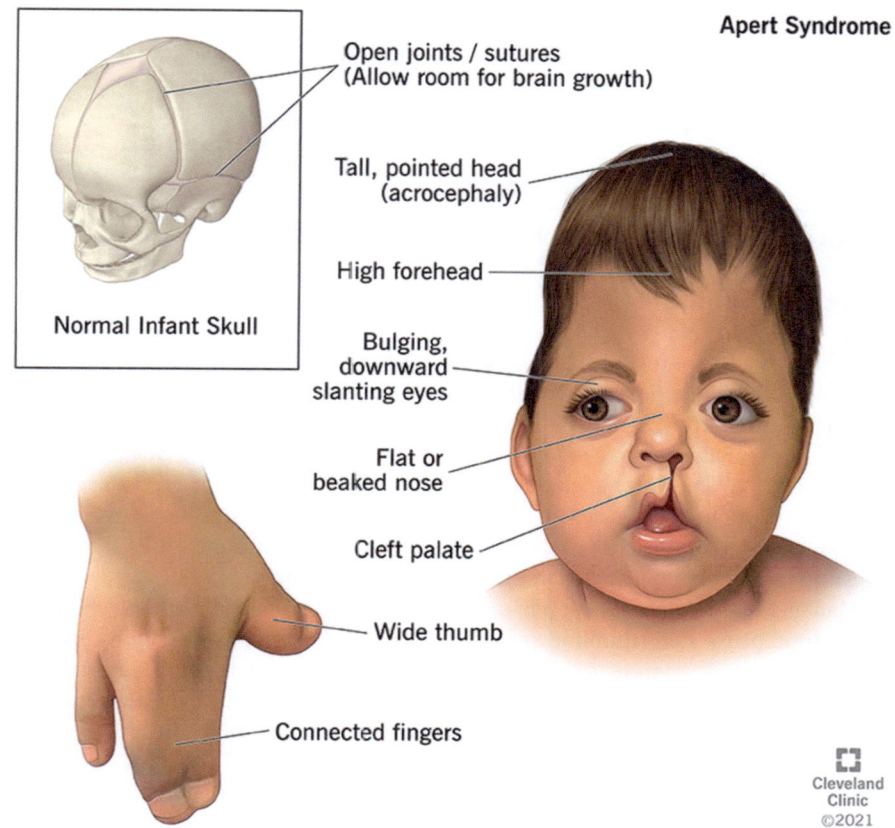

Fig. 16.3 Apert syndrome airway anatomy

Management: Treatment for sleep disorders in Prader-Willi Syndrome (PWS) frequently includes continuous positive airway pressure (CPAP) therapy to manage OSA. For EDS, medications like modafinil may be used to enhance wakefulness. Additionally, melatonin and rigorous sleep hygiene practices are commonly implemented to address disruptions in circadian rhythms [22, 23].

Duchenne Muscular Dystrophy (DMD)

Duchenne muscular dystrophy (DMD) is a condition marked by neuromuscular hypotonia, which significantly impacts respiratory function and sleep. The progressive muscle weakness inherent in DMD leads to a decline in lung capacity and contributes to upper airway obstruction, both of which are significant factors in the development of OSA. Studies have shown that up to 90% of patients with DMD experience SDB, including OSA, due to these respiratory complications [24, 25].

Additionally, respiratory muscle weakness in DMD can exacerbate daytime hypoventilation and contribute to sleep-related breathing disturbances [26, 27].

Management: Non-invasive ventilation techniques, such as bilevel positive airway pressure (BiPAP), are essential in managing SDB in DMD. Pulmonary rehabilitation and early intervention with respiratory therapies are also crucial [25–27].

Syndromes with Neurodevelopmental Dysfunction

Angelman Syndrome (AS)

Angelman syndrome results from mutations or deletions of the UBE3A gene, which leads to profound intellectual disability, motor dysfunction, and epilepsy. Sleep disturbances affect up to 80% of children with AS, among which insomnia and frequent nocturnal awakenings are the most common complaints [28]. Several studies report abnormal sleep architecture with reduced REM sleep, prolonged sleep onset latency, and fragmented sleep patterns [29]. Dysfunction in the GABA Gamma-Aminobutyric Acid System (GABAS) plays a key role in malfunction of the neurotransmission, likely contributes to the sleep problems seen in AS [30].

Management: Melatonin is often used to manage sleep onset insomnia which has shown to improve sleep duration and quality [31]. Behavioral interventions, such as consistent bedtime routine and limiting stimulating activities before sleep, are also effective. In some cases, antiepileptic drugs such as clonazepam may be prescribed to reduce nocturnal awakenings, especially in children with concurrent epilepsy [32].

Fragile X Syndrome (FXS)

Fragile X syndrome is the most common inherited cause of intellectual disability and is frequently associated with autism spectrum disorder (ASD). Insomnia, frequent nocturnal awakenings, and restless sleep are reported in 50–70% of children with FXS [33]. Anxiety, hyperarousal, and sensory sensitivities commonly exacerbate these sleep problems.

Children with FXS also show abnormalities in sleep architecture, such as reduced REM sleep and frequent transitions between sleep stages. These issues contribute to daytime behavioral challenges, including hyperactivity, mood instability, and irritability [34].

Management: Behavioral sleep interventions, including maintaining a structured bedtime routine and reducing environmental stimuli, are essential for managing sleep problems in FXS. Melatonin supplementation is commonly used, and medications such as clonidine may help with sleep initiation. However, these pharmacological interventions should be used cautiously, due to the potential side effects [33, 34].

Rett Syndrome

Rett syndrome, primarily affecting girls, is caused by mutations in the MECP2 gene. Children with Rett syndrome often experience SDB (OSA, CSA) and circadian rhythm disturbances. Affected children may show irregular sleep patterns, with frequent night-time awakenings and reduced overall sleep time [35].

Management: Like other neurodevelopmental disorders, behavioral strategies such as maintaining a structured sleep routine pharmacologic therapies like melatonin and medications targeting respiratory function, such as CPAP for OSA, are used [36].

Syndromes with Multi-system Involvement

Smith-Magenis Syndrome (SMS)

Smith-Magenis syndrome (SMS) is caused by a deletion or mutation on chromosome 17p11.2 and is notorious to cause severe circadian rhythm disturbances. These children exhibit a reversed melatonin secretion pattern, where melatonin levels are elevated during the day and reduced at night, leading to daytime sleepiness and nocturnal wakefulness [37]. Psychosocial issues, including self-injury and aggression, often worsen due to chronic sleep deprivation [38].

Management: Circadian rhythm disturbances in SMS are typically treated with beta-blockers like propranolol to suppress daytime melatonin and evening melatonin supplementation to promote night-time sleep. Behavioral therapies that reinforce consistent sleep-wake cycles are also essential in improving sleep patterns [37–39].

Clinical Pearls for Managing Sleep Disturbances in Genetic Syndromes

1. Early Detection and Multidisciplinary Approach: Early sleep studies are crucial in identifying sleep disorders like OSA in conditions such as Down syndrome and PRS. A multidisciplinary approach, including pulmonologists, ENT surgeons, and sleep specialists, is often required for effective management.
2. Airway Management: Prone positioning can help infants with PRS improve airway patency. For severe cases, surgical interventions like mandibular distraction may be necessary. Children with conditions like Apert syndrome often require surgeries like adenotonsillectomy and midface advancement to address airway obstruction.
3. CPAP and Non-invasive Ventilation: CPAP is essential for managing OSA in conditions with hypotonia like Down syndrome and Prader-Willi syndrome (PWS). It helps prevent airway collapse during sleep. Children with Duchenne muscular dystrophy benefit from BiPAP to support respiratory function and manage SDB.
4. Managing Circadian Rhythm Disruptions: Melatonin is used in neurodevelopmental syndromes like Angelman and Fragile X syndromes to improve sleep

onset and duration. In Smith-Magenis syndrome, beta-blockers and melatonin help correct reversed sleep patterns.
5. Behavioral and Pharmacological Interventions: Establishing consistent sleep routines is vital, particularly for Rett and Fragile X syndromes. Medications like modafinil and melatonin can manage excessive daytime sleepiness (EDS) and circadian rhythm disruptions in Prader-Willi syndrome.
6. Long-Term Monitoring: Regular sleep studies are essential to assess treatment effectiveness. Ongoing monitoring helps adjust CPAP settings and evaluate surgical outcomes as the child grows.

Conclusion

Sleep disturbances are a significant concern in children with genetic syndromes, driven by a variety of anatomical, neuromuscular, and neurodevelopmental abnormalities. These sleep problems exacerbate the clinical manifestations of the syndromes and negatively affect quality of life. A multidisciplinary approach, including behavioral, medical, and sometimes surgical interventions, is essential for optimizing outcomes and enhancing the overall well-being of these patient populations.

References

1. Blackmer AB, Feinstein JA. Management of sleep disorders in children with neurodevelopmental disorders: a review. Pharmacotherapy. 2016;36(1):84–98. https://doi.org/10.1002/phar.1686.
2. Robinson-Shelton A, Malow BA. Sleep disturbances in neurodevelopmental disorders. Curr Psychiatry Rep. 2016;18(6):6. https://doi.org/10.1007/s11920-015-0638-1.
3. Marincak Vrankova Z, Krivanek J, Danek Z, et al. Candidate genes for obstructive sleep apnea in non-syndromic children with craniofacial dysmorphisms: a narrative review. Front Pediatr. 2023;11:1117493. https://doi.org/10.3389/fped.2023.1117493. Published June 27, 2023
4. Shott SR, Amin R, Chini B, et al. Obstructive sleep apnea: should all children with down syndrome be tested? Arch Otolaryngol Head Neck Surg. 2006;132(4):432–6.
5. De Miguel-Díez J, Villa-Asensi JR, Álvarez-Sala JL. Prevalence of sleep-disordered breathing in children with down syndrome: Polygraphic findings in 108 children. Sleep. 2003;26(8):1006–9.
6. Churchill SS, Kieckhefer GM, Bjornson KF, Herting JR. Relationship between sleep disturbance and functional outcomes in daily life habits of children with down syndrome. J Dev Behav Pediatr. 2013;34(2):97–105.
7. Mitchell RB, Kelly J. Outcome of adenotonsillectomy for obstructive sleep apnea in children under 3 years. Arch Otolaryngol Head Neck Surg. 2005;131(2):95–100.
8. Lin HY, et al. Dental and orofacial anomalies in children with Prader-Willi syndrome. J Clin Pediatr Dent. 2010;34(3):275–9.
9. Dell Children's Craniofacial Team of Texas. Pierre Robin Sequence (PRS). Published March 4, 2017. Accessed 19 Nov 2024. https://craniofacialteamtexas.com/pierre-robin-sequence-prs/
10. Doudney K, Stanier P. Epithelial cell polarity genes are required for neural tube closure. Am J Med Genet C Semin Med Genet. 2005;135C(1):42–7. https://doi.org/10.1002/ajmg.c.30052.

11. Cohen SM, Greathouse ST, Rabbani CC, et al. Robin sequence: What the multidisciplinary approach can do. J Multidiscip Healthc. 2017;10:121–32. https://doi.org/10.2147/JMDH.S98967. Published March 27, 2017
12. Shapiro EG, et al. Sleep-disordered breathing in children with achondroplasia. Pediatrics. 2017;140(2):e20170109. https://doi.org/10.1542/peds.2017-0109.
13. Tasker RC, Cole GF. Central sleep apnea in achondroplasia: mechanisms and outcomes. J Child Neurol. 2018;33(4):299–305. https://doi.org/10.1177/0883073818757021.
14. Thompson JR, et al. Clinical outcomes of decompression surgery for foramen magnum stenosis in achondroplasia. J Neurosurg Pediatr. 2019;24(6):635–44. https://doi.org/10.3171/2019.6.PEDS19512.
15. Wright MJ, Irving MD. Clinical management of achondroplasia: current perspectives. Ther Clin Risk Manag. 2012;8:187–99. https://doi.org/10.2147/TCRM.S25600.
16. Trotter TL, Hall JG. Health supervision for children with achondroplasia. Pediatrics. 2005;116(3):771–83. https://doi.org/10.1542/peds.2005-1440.
17. Lee SK, et al. Sleep and airway assessment in children with Apert syndrome. J Craniofac Surg. 2012;23(6):1674–7.
18. Cleveland Clinic. Apert syndrome: what is it, symptoms, diagnosis & treatment. Published May 1, 2024. https://my.clevelandclinic.org/health/diseases/22077-apert-syndrome
19. Thompson NM, et al. Upper airway obstruction in Apert's syndrome: treatments and outcomes. Cleft Palate Craniofac J. 1995;32(3):189–94.
20. Duvall RS, Williams RJ. Prevalence of obstructive and central sleep apnea in Prader-Willi syndrome. J Clin Sleep Med. 2012;8(5):497–505. https://doi.org/10.5664/jcsm.2226.
21. Ingram DG, Arganbright JM, Halpin KL, Okorie C. Sleep challenges in children with Prader-Willi syndrome: a patient and family handout. ATS Scholar. 2021;2(4):665–5681. https://doi.org/10.34197/ats-scholar.2021-0025PE. Published December 17, 2021
22. Williams JM, Eastwood PR. Sleep-disordered breathing in Prader-Willi syndrome: pathophysiology and management. Sleep Med Rev. 2010;14(2):81–7. https://doi.org/10.1016/j.smrv.2009.05.002.
23. Horne RS, McBride J. Sleep-related respiratory disorders in genetic syndromes. Clin Pulm Med. 2014;21(4):194–204. https://doi.org/10.1097/CPM.0000000000000039.
24. Montagnese S, Pinna GD. Respiratory and sleep disorders in Duchenne muscular dystrophy. Muscle Nerve. 2017;56(2):223–31. https://doi.org/10.1002/mus.25464.
25. McDermott MP, Abresch RT. Prevalence and impact of sleep disorders in Duchenne muscular dystrophy. Am J Respir Crit Care Med. 2008;178(4):431–8. https://doi.org/10.1164/rccm.200803-410OC.
26. D'Angelo MG, Bianchi M. Respiratory impairment and sleep disorders in DMD: clinical insights and management strategies. Neuromuscul Disord. 2015;25(6):503–11. https://doi.org/10.1016/j.nmd.2015.03.002.
27. Mazzone ES, Catterall A. Assessment and management of sleep-disordered breathing in Duchenne muscular dystrophy. Sleep Med Rev. 2016;25:22–30. https://doi.org/10.1016/j.smrv.2015.04.002.
28. Williams CA, Beaudet AL. Angelman syndrome: clinical and genetic aspects. J Child Neurol. 2018;33(8):561–9. https://doi.org/10.1177/0883073818763387.
29. de Weerd A, Overeem S. Sleep architecture and sleep disorders in Angelman syndrome. Sleep Med Rev. 2006;10(2):163–74. https://doi.org/10.1016/j.smrv.2005.07.002.
30. Hines RM. GABAergic dysfunction in Angelman syndrome. The role of GABA in sleep disorders associated with Angelman syndrome. Front Neurol. 2014;5:205. https://doi.org/10.3389/fneur.2014.00205.
31. Baker BL, Richards S. Melatonin for sleep problems in Angelman syndrome. J Child Neurol. 2008;23(11):1342–6. https://doi.org/10.1177/0883073807307331.
32. Nielsen JE, Egeberg KT. Melatonin treatment for sleep problems in Angelman syndrome: a systematic review. Sleep Med Rev. 2013;17(5):369–74. https://doi.org/10.1016/j.smrv.2012.08.003.

33. Hirsch A, Eide M. Sleep problems in fragile X syndrome: a review of the literature. J Dev Behav Pediatr. 2006;27(2):148–55. https://doi.org/10.1097/00004703-200604000-00012.
34. Carter CS, Cote KA. Sleep and behavioral disturbances in fragile X syndrome: a review of the literature. J Dev Behav Pediatr. 2009;30(1):70–80. https://doi.org/10.1097/DBP.0b013e3181904e3b.
35. Katz DJ, Neul JL. Sleep disturbances in Rett syndrome: a review. Curr Opin Neurol. 2013;26(2):183–90. https://doi.org/10.1097/WCO.0b013e32835b7f4a.
36. Smith AC, McGavran L. Smith-Magenis syndrome: clinical and genetic features. Am J Med Genet. 2005;137C(1):55–60. https://doi.org/10.1002/ajmg.c.30052.
37. Miller JC, Morris CA. Psychosocial challenges and sleep disturbances in Smith-Magenis syndrome: impact and management. J Intellect Disabil Res. 2012;56(7):755–64. https://doi.org/10.1111/j.1365-2788.2012.01583.x.
38. Smith AC, Theisen HM. Treatment of circadian rhythm disorders in Smith-Magenis syndrome using propranolol. Am J Med Genet C Semin Med Genet. 2003;120C(1):66–71. https://doi.org/10.1002/ajmg.c.20009.
39. Bowers AW, Magenis E. Propranolol for circadian rhythm disorders in Smith-Magenis syndrome: a pilot study. Sleep Med Rev. 2006;10(5):353–60. https://doi.org/10.1016/j.smrv.2005.09.003.

Correction to: Sleep, Chronotype, and Learning: A Developmental Perspective

Marty Martin and Zeeshawn Malik

Correction to:
Chapter 3 in: A. Wani, I. S. Khawaja (eds.), *Sleep Disorders in Children*, https://doi.org/10.1007/978-3-031-92166-7_3

This chapter was recently published. However, the authorship was listed incorrectly and has been corrected in the revised publication, as shown below.

Current Author Order

Marty Martin - Department of Management & Entrepreneurship, Driehaus College of Business, DePaul University, Chicago, IL, USA
e-mail: martym@depaul.edu

Updated Author Order with Affiliations

Marty Martin - Department of Management & Entrepreneurship, Driehaus College of Business, DePaul University, Chicago, IL, USA
e-mail: martym@depaul.edu

Zeeshawn Malik - The University of Texas at Dallas, Richardson, TX, USA

The updated version of this chapter can be found at:
https://doi.org/10.1007/978-3-031-92166-7_3

Index

A
Academic performance, 6–7, 24, 26–29, 31
Achondroplasia, 169, 170
Actigraphy, 135, 136
Adenotonsillectomy, 88, 168
Adolescent brain maturation, 74
Adolescent sleep, 34
    brain maturation, 74
    cultural and socioeconomic factors, 79
    developmental and biological changes, 73–74
    DSWPD, 75
    health and well-being, 79
    insomnia, 76
    OSA, 77
    parasomnias, 77
    physiology of, 71–73
    psychosocial and behavioral factors, 78
    sleep regulation, 74
Adolescents, 78, 79
Aerophagia, 88
Allergies, 55
Angelman syndrome (AS), 172
Antidepressants, 162
Antiepileptic therapies, sleep and, 149
Anxiety, 36, 140
Apert syndrome, 87, 170, 171
Apnea-hypopnea index (AHI), 87, 88
Apneas, 96
Atopic dermatitis, 56
Attention-deficit/hyperactivity disorder (ADHD), 140, 164
Autism spectrum disorder (ASD), 153, 154, 172

B
Basal ganglia, 113
BEARS questionnaire, 64, 65
Bedtime routine, 8–9, 46
Behavior effects, 6–7
Benzodiazepines, 115
Bilevel positive airway pressure (BiPAP), 88, 97
Bilevel positive airway pressure with spontaneous-timed mode (BiPAP-ST), 97
Bright light exposure, 17–19
Brodsky grading scale of tonsil size, 86
Bruxism, 124
    clinical and laboratory evaluation, 125
    consequences, 126
    diagnosis, 125
    differential diagnosis, 125
    pathophysiology, 124
        central nervous system involvement, 124
        dental factors, 124
        genetic predisposition, 124
        psychological factors, 124
        sleep disorders, 124
    treatment, 126

C
Cataplexy, 138
Central apnea index (CAI), 95, 98
Central disorder of hypersomnolence (CDH), 138
Central pattern generators (CPG), 113

Central sleep apnea (CSA), 85, 95, 96, 149
Cerebral palsy (CP), 151–153, 156
Chiari malformations, 95
Chiari type 1 malformation, 95
Children and adolescents, 3
Children's Sleep Habits Questionnaire (CSHQ), 64
Chronic sleep deprivation, 6
Chronotherapy, 75
Chronotype, 23–25, 27–31, 34
Circadian rhythm, 34, 35, 71–73, 75
Circadian rhythm disturbances, 173
Circadian rhythm sleep-wake disorder (CRSWD), 65, 67
Clonazepam, 172
Cognitive behavioral therapy for insomnia (CBT-I), 31, 76
Cognitive development, 24
Cognitive functioning, 6–7
Cognitive impairments, 6–7
Complex sleep apnea (CompSA), 96
Congenital central hypoventilation syndrome (CCHS), 95, 96
Continuous positive airway pressure (CPAP) therapy, 77, 88, 96–98, 126
Craniofacial and airway abnormalities
   achondroplasia, 169, 170
   Apert syndrome and, 170
   Down syndrome, 168, 169
   Pierre Robin Sequence, 169

D
Delayed sleep phase syndrome (DSPS), 67, 162
Delayed sleep-wake phase disorder (DSWPD), 18, 75
Depression, 36, 140, 164
Desmopressin, 108
Diabetes, 137
Dim light melatonin onset (DLMO), 34
Disturbed sleep, 164
Down syndrome (DS), 139, 168–169
Drug-induced sleep endoscopy (DISE), 87
Duchenne muscular dystrophy (DMD), 87, 171
Dysregulation, 35

E
Elementary school children, 33
Emotional dysregulation, 164
Epilepsy
   interface of, 147
   sleep and antiepileptic therapies, 149
   sleep disorders in, 148, 149
   sleep related epilepsy disorders, 149
Epileptic Encephalopathy with Spike Wave Activation in Sleep (EE-SWAS), 149
Epworth Sleepiness Scale (ESS), 31
Excessive daytime sleepiness (EDS), 131
   behavioral/mental health conditions, 140
   daytime fatigue, 131
   differential diagnosis, 137–139
   evaluation, 134
   exam, 136
   examination/mental status exam, 133
   incidence, 132
   interpretation, 134–136
   paradoxical symptoms, 131
   questions on, 134
   respiratory system related causes, 139
   symptoms of, 132
Excessive sleepiness, 78
Extreme prematurity, 139

F
Fear of missing out (FOMO), 78
Food allergy, 140
Fragile X syndrome (FXS), 172

G
Gabapentin, 149
Gabapentinoids, 149
Gaming consoles, 17
Gastroesophageal reflux (GERD), 152
Generalized anxiety disorder (GAD), 164
Genetic syndromes, 139, 167
   with craniofacial and airway abnormalities
      achondroplasia, 169, 170
      Apert syndrome and, 170
      Down syndrome, 168, 169
      Pierre Robin Sequence, 169
   with multi-system involvement
      Smith-Magenis syndrome, 173
   with neurodevelopmental dysfunction
      Angelman syndrome, 172
      fragile X syndrome, 172
      Rett syndrome, 173
   with neuromuscular hypotonia
      Duchenne muscular dystrophy, 171, 172
      Prader-Willi syndrome, 170, 171
   sleep disturbances in, 173
Ghrelin, 35
Growth hormones, 5

H

Headache
  and sleep interface, 150
  sleep disorders in, 150, 151
Health equity, 23, 28, 31
Healthy sleep duration, 34
Healthy sleep public policy, 38
Heated high-flow nasal cannula oxygen, 90
Hemoglobinuria, 115
High flow nasal cannula (HFNC), 89
Hypersomnia, 149
Hypocretin, 138
Hypothalamic suprachiasmatic nucleus, 4
Hypothyroidism, 137

I

Idiopathic hypersomnia (IH), 138
Image rehearsal therapy (IRT), 108
Inflammatory cytokines, 36
Insomnia, 52, 76, 78
Interictal epileptiform, 147
Interictal epileptiform discharges (IEDs), 149
Interleukin-1 (IL-1), 36
International RLS Study Group (IRLSSG), 116
Internet addiction, 140
Intranasal corticosteroids for OSA, 88
Iron deficiency, 116, 138

K

Klein-Levin syndrome (KLS), 139

L

Learning, 23–31
  adults, 30, 31
  college/university, 29–30
  elementary school, 26
  high school, 28, 29
  middle school, 27, 28
  preschool and kindergarten, 24–26
Leptin, 35
Leukotriene inhibitors, 88

M

Medical insomnia, 55–56
Melatonin, 153, 172
Memory, 23–31
Mental health, 36, 140, 162
  issues, 9
  sleep and, 163
  biological mechanisms linking sleep and mental health, 164
  impact of, 164
  interventions and therapeutic approaches, 163
  sleep deprivation and emotional dysregulation, 164
Middle schoolers, 37
Midface hypoplasia, 88
Mixed apnea (MA), 97
Multiple Sleep Latency Test (MSLT), 135, 161
Multi-system involvement, Smith-Magenis syndrome, 173
Myotonic Dystrophy (MD), 139

N

Narcolepsy, 138, 149
Neurodevelopmental dysfunction
  Angelman syndrome, 172
  fragile X syndrome, 172
  Rett syndrome, 173
Neurological disorders, 156
  differential diagnosis, 156
  pediatric patients, 157
  plan, 156
  and sleep, 147
    autism spectrum disorder, 153, 154
    cerebral palsy, 151–153
    diagnosis and treatment, considerations for, 154, 155
    headaches, 150, 151
    seizures and epilepsy, 147–149
Neuromuscular hypotonia
  Duchenne muscular dystrophy, 171, 172
  Prader-Willi syndrome, 170, 171
Newborn sleep patterns, 43–44
Night terrors, 103
Night wakings, 45, 48
Nightmare disorder, 104, 108
Nightmares, 107
Nighttime awakenings, 52
Nocturnal enuresis, 108
Nocturnal frontal lobe epilepsy (NFLE), 149
Non-REM (NREM) sleep, 36, 71
Non-REM related parasomnias, 101, 103
Normal sleep, 52

O

Obesity, 35
Obstructive sleep apnea (OSA), 54, 65, 77, 85, 139, 152, 156, 162
  clinical presentation and exam findings, 86

Obstructive sleep apnea (OSA) (*cont.*)
  diagnosis, 87
  prognosis, 89
  risk factors, 86, 87
  treatment, 88, 89
Optimal sleep duration, 4
Oxazepam, 115

P
Parasomnias, 65, 77, 101, 149
  diagnosis and evaluation, 105, 106
  differential diagnosis, 106
  follow-up, 109
  management and treatment, 106, 107
    nightmare disorder, 108
    nocturnal enuresis, 108
    sleep terrors, 107, 108
  pathophysiology, 105
  types of, 101, 102
    non-REM, 101, 103, 104
    sleep related hallucinations, 104
    sleep-related urologic dysfunction, 104
Pediatric insomnia, 52
Pediatric obesity, 35
Pediatric sleep, 20, 21
Pediatric sleep disorders, 65–66
Pediatric Sleep Questionnaire (PSQ), 152
Pediatric Sleep-Questionnaire—Sleep-Related Breathing Disorders (PSQ-SRBD) Scale, 87
Periodic breathing, 96
Periodic leg movement disorder (PLMS), 122, 123
Periodic limb movement disorder (PLMD), 56, 65, 152
Pierre Robin Sequence (PRS), 87, 169
PLMS index (PLMI), 122
Polysomnogram, 136
Polysomnography (PSG), 95, 111, 135, 161
Poor sleep, 78
Positional therapy, 89
Positive airway pressure (PAP) therapy, 88, 90, 153
Post-puberty, 87
Prader-Willi syndrome (PWS), 139, 170, 171
Pregabalin, 149
Prematurity, 87
Primary care, 23, 25, 26, 30, 31
Primary care providers, 64–65
Psychological stimulation, 18

R
Rapid eye movement (REM) sleep, 43, 72
Recurrent isolated sleep paralysis, 104
REM related parasomnias, 103, 104
Respiratory conditions, 56
Restless legs syndrome (RLS), 56–57, 65, 118, 138, 162, 165
  assessment, 161
  clinical and laboratory evaluation, 121
  clinical features, 119
  consequences, 121
  diagnosis, 119, 120
  differential diagnosis, 121
  pathophysiology, 119
  treatment, 121, 122
Restless sleep disorder (RSD), 57, 65, 115, 116
  clinical and laboratory evaluation, 117, 118
  diagnosis, 117
  pathophysiology, 116
    iron deficiency, 116
    sleep instability, 116, 117
    sympathetic activation, 117
  therapeutic options, 118
Rett syndrome, 173
Rhythmic movement disorder, 114

S
School aged children sleep
  barriers, 64
  differential diagnosis, 67
  history, 66
  optimal sleeping environment, 63
  physical examination, 66
  testing/studies, 66
School athletes, 36
School start times, 71, 74, 78
  benefits of, 37
  public health policy, 37–38
Screen time
  impact on sleep, 19
  near bedtime, 17
  outcomes, 18–19
  recommendations, 20–21
Secondhand smoke exposure (SHS), 87
Seizures, 147
Self-limited Epilepsy with Centrotemporal Spikes (SeLECTS), 149
Self-soothing, 45–47, 112
Separation anxiety, 45, 47

Index

Short sleep duration, 7
Shorter sleep duration, 19
Skin irritation, 88
Sleep
    and antiepileptic therapies, 149
    architecture, 71
    and mental health, 163
        biological mechanisms linking sleep and mental health, 164
        impact of, 164
        interventions and therapeutic approaches, 163
        sleep deprivation and emotional dysregulation, 164
    neurological disorders and, 147
        autism spectrum disorder, 153, 154
        cerebral palsy, 151–153
        diagnosis and treatment, considerations for, 154, 155
        headaches, 150, 151
        seizures and epilepsy, 147–149
    and wakefulness, 4
Sleep cycles, 44
Sleep deprivation, 74, 80, 162, 164
    assessment, 12
    diagnosis and assessment, 5–6
    examination, 11
    medication, 11
    past medical history, 10
    prevention strategies, 7–9
    screening, 11
    sleep history, 10
    social history, 10
    treatment for, 8
Sleep development, 45
Sleep diary, 12
Sleep disordered breathing (SDB), 54
Sleep disorders
    in CP, 152, 153
    in epilepsy, 148, 149
    examination/mental status exam, 161
    follow-up, 163
    in headache, 150, 151
    treatment, 163
Sleep Disorders Inventory for Students-Revised-Adolescent (SDIS-R-A), 12
Sleep displacement, 18
Sleep Disturbance Scale for Children (SDSC), 12
Sleep disturbances, 167
Sleep duration, 3, 5, 51
Sleep environment, 45, 46, 48
Sleep hygiene, 74, 75, 77
Sleep hygiene education, 31
Sleep instability, 112–113, 115–117
Sleep interface, 150
Sleep issues, 47
Sleep patterns, 78, 164
Sleep quality
    and duration, 38
    and quantity, 36
Sleep regressions, 45
Sleep related epilepsy disorders, 149
Sleep related hallucinations, 104
Sleep screening tools, 64–65
Sleep terrors, 103, 107, 108
Sleep-friendly environment, 45
Sleep-related eating disorder (SRED), 103
Sleep-related hypoventilation, 85
Sleep-related movement disorders, 111, 112
    bruxism, 124
        clinical and laboratory evaluation, 125
        consequences, 126
        diagnosis, 125
        differential diagnosis, 125
        pathophysiology, 124
        treatment, 126
    periodic leg movement disorder, 122, 123
    restless legs syndrome, 118
        clinical and laboratory evaluation, 121
        clinical features, 119
        consequences, 121
        diagnosis, 119, 120
        differential diagnosis, 121
        pathophysiology, 119
        treatment, 121, 122
    restless sleep disorder, 115, 116
        clinical and laboratory evaluation, 117, 118
        diagnosis, 117
        pathophysiology, 116, 117
        therapeutic options, 118
    sleep-related rhythmic movement disorders, 112
        clinical and laboratory evaluation, 114, 115
        clinical features and comorbidities, 113
        diagnosis, 113, 114
        differential diagnosis, 114
        pathophysiology, 112, 113
        prescription medication, 115
        treatment options, 115
Sleep-related rhythmic movement disorders (SRRMD), 112
    clinical and laboratory evaluation, 114, 115
    clinical features and comorbidities, 113
    diagnosis, 113, 114

Sleep-related rhythmic movement disorders
    (SRRMD) (cont.)
  differential diagnosis, 114
  pathophysiology, 112
    central pattern generators, 113
    genetics, 113
    self-soothing, 112
    sleep instability, 112
  prescription medication, 115
  treatment options, 115
Sleep-wake cycle, 19, 74
Sleepwalking, 77, 103
Slow-wave sleep, 72
Smartphones, 17
Smith-Magenis syndrome (SMS), 173
Social Determinants of Health, 54
Social jet lag, 27, 71
Socioeconomic status (SES), 79
Standford Sleepiness Scale (SSS), 135
Strategic timing, 75
Sympathetic activation, 117
Synaptic pruning, 74

T
Tablets, 17
Teenagers, 71, 78

Teeth grinding, 124
Televisions, 17
Timed wakings, 77
Toddler sleep
  assessment, 53–54
  clinical consequences, 59–60
  diagnosis, 53–54
  differential diagnosis, 54–57
  etiology, 53
  insomnia, 52
  pharmacological management, 59
  SDB, 54
  Social Determinants of Health, 54
  treatment, 58–59
  workup, 57–58
Tonsillectomy and adenoidectomy (T&A), 97
Trisomy 21, 168–169

U
Unhealthy sleep, 109
Upper airway infections, 55
US Food and Drug Administration (FDA), 59

V
Vagus nerve stimulation (VNS), 149

If you have any concerns about our products,
you can contact us on
ProductSafety@springernature.com

In case Publisher is established outside the EU,
the EU authorized representative is:
**Springer Nature Customer Service Center GmbH
Europaplatz 3, 69115 Heidelberg, Germany**

Printed by Libri Plureos GmbH
in Hamburg, Germany